Changes in Latitude

An Uncommon Anthropology

Joana McIntyre Varawa

photographs by the author

Harper & Row, Publishers, New York
Grand Rapids, Philadelphia, St. Louis, San Francisco
London, Singapore, Sydney, Tokyo, Toronto

A hardcover edition of this book was published in 1989 by The Atlantic Monthly Press. It is here reprinted by arrangement with The Atlantic Monthly Press.

For information address Harper & Row, Publishers, Inc., 10 East 53rd Street, New York, N.Y. 10022.

First PERENNIAL LIBRARY edition published 1990.

LIBRARY OF CONGRESS CATALOG CARD NUMBER 89-46126

ISBN 0-06-097319-6

91 92 93 94 FG 10 9 8 7 6 5 4 3 2

Praise for *Changes in Latitude*

"Varawa is an accomplished writer, who lovingly describes the tropical paradise she has come to live in. A surprisingly affecting, if unconventional story."

—*Washington Post Book World*

"By turns touching, astonishing, amusing, and deeply human. . . . Joana Varawa writes with such empathy . . . and lyrical fervor of natural splendors and horror alike, that the book is utterly disarming."

—*Publishers Weekly*

"A vivid account."

—*New York Times Book Review*

"Varawa is a refugee from modernity and a seeker of the ecological grail of communal, self-sufficient life. . . . *Changes in Latitude* is worth reading for her close-up look at Fijian culture. Its enchantments are many, but the author's wish to live up to her fantasy is stronger still."

—*Newsday*

"Imagine switching roles, and lives, with someone like Gauguin. That's the enchantment of Joana McIntyre Varawa's *Changes in Latitude,* an offbeat memoir that chronicles the author's sudden move to the Fiji islands and the astonishing marriage proposal that brought her there."

—*Glamour*

"[Varawa] pays tribute to the beautiful complexity of life among [her] own species."

—*San Jose Mercury News*

ACKNOWLEDGMENTS

I would like to express my deep gratitude to the following people, whose help and advice made this book possible: Maleli Varawa, James and Carolyn Robertson, Ann Godoff, Navi Taga, Barbara Youngblood.

FIJI ISLANDS

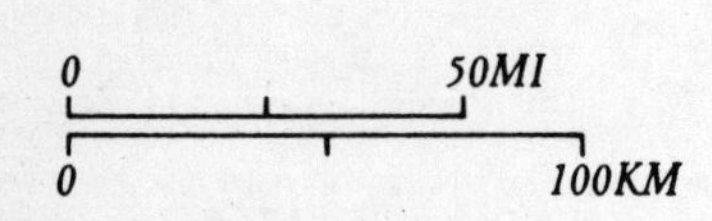

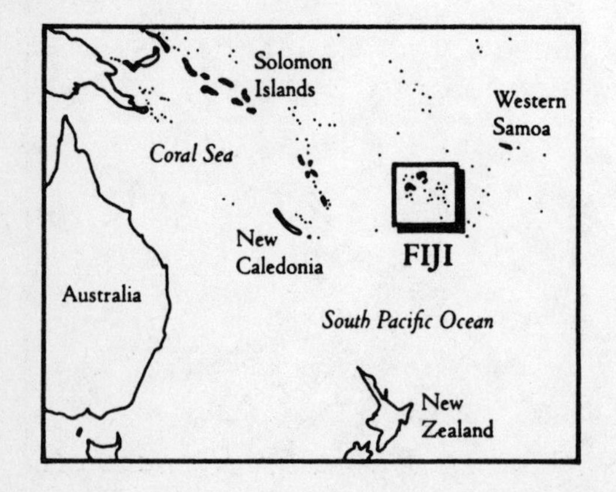

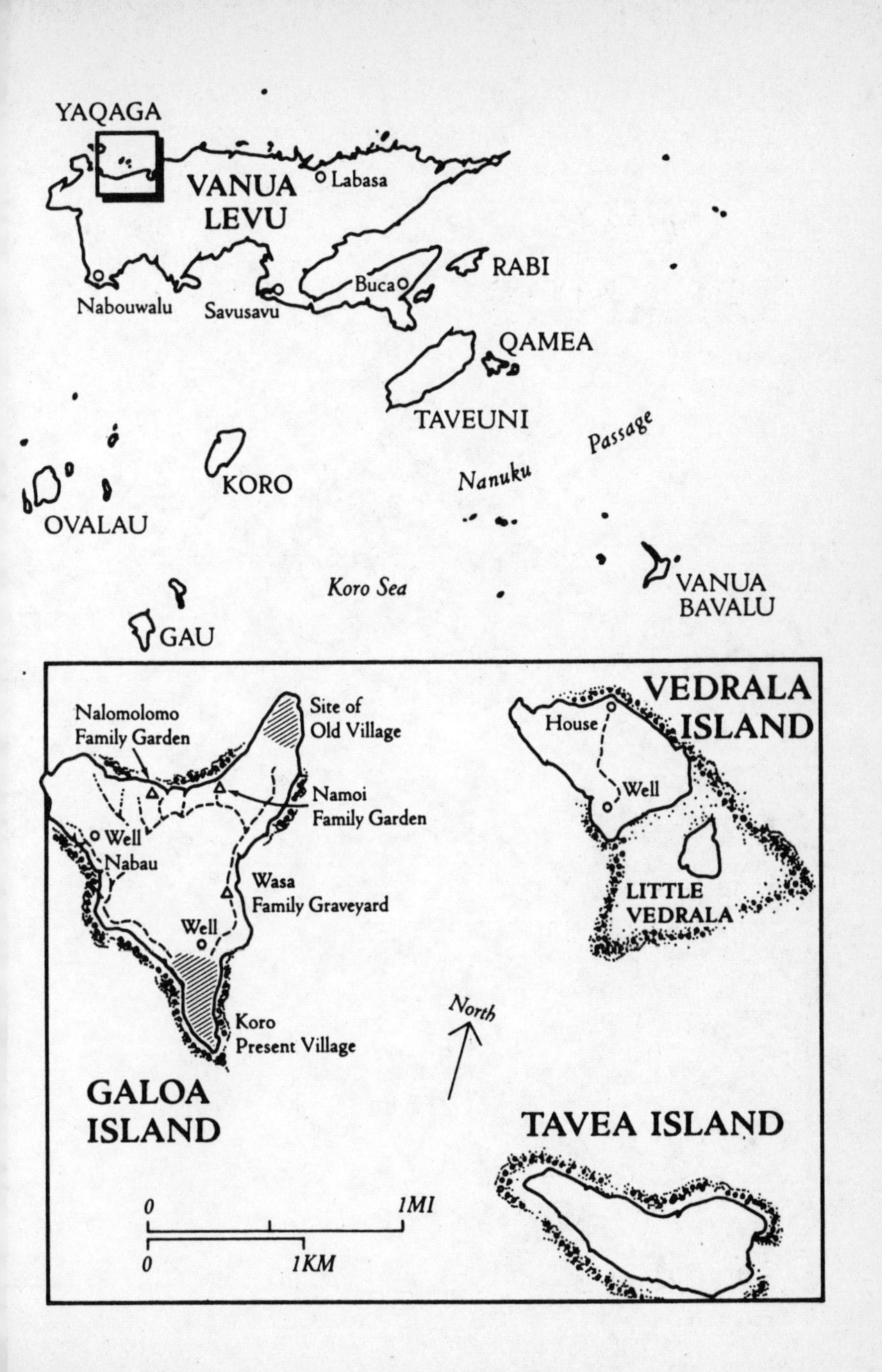

YAQAGA
VANUA LEVU
Labasa
RABI
Buca
Nabouwalu
Savusavu
QAMEA
TAVEUNI
Passage
Nanuku
KORO
OVALAU
Koro Sea
VANUA BAVALU
GAU
VEDRALA ISLAND
House
Well
LITTLE VEDRALA
Nalomolomo Family Garden
Site of Old Village
Namoi Family Garden
Well
Nabau
Wasa Family Graveyard
Well
Koro Present Village
North
GALOA ISLAND
TAVEA ISLAND
0
1MI
0
1KM

CONTENTS

CHANGES IN LATITUDE

THE MEETING

A succulent orange moon lies close to the horizon, flooding the sea with golden light. Somewhere far below us an invisible line changes today into tomorrow, and somewhere behind us now a hurricane whips the sea into frothy torments. Inside the cabin of the plane, all is still, my fellow passengers drifting in the uneasy somnolence of midnight flight. Soon we will land.

I am on my way to Fiji from Honolulu, and I am looking for something, something bushy and wild. When I was young, my mother was forever telling me to comb my hair because I looked like a Fijian. The word came to stand for all that was exotic and forbidden to a small awkward girl growing up in the sticky asphalt streets of Chicago and Los Angeles. I never did manage to tame my hair, nor my mind. The primitive still beckons, though I haven't the faintest idea what reality lies behind that fantasy.

Inside my purse is a postcard. On the front of the card is a photograph of a big schooner coursing through a pale blue sea. On the back are a few scrawled words from my friend Homer, who has just gone to Fiji to work as the captain of the schooner. He writes: "Come, you would love it here. The old ways still exist, even in the cities."

His card reminds me of the one Fijian I have met, a boat builder named Sam Whippy, who repaired my sailboat a few years ago in Honolulu. Sam captivated me by his calm competence, his charm, and his stories of village life in Fiji. He described a place I imagined as romantically remote, lush green, and scarcely inhabited, a place where I could go and stay with one of his many uncles and learn the Fijian way. Sam talked about villages where the men sat all night drinking *yaqona*, the universal South Pacific tranquilizer, of the soft music and the stories, of villages of timeless ease, and of smiles that welcomed strangers. Sam's stories nourished my fantasies.

I was born in California, and after pursuing a career and having a family there, I felt myself drawn west. It seemed to be my direction, so I moved to Hawaii. After spending ten years there, I was ready to go even farther west. It is not easy in Hawaii to live simply and consonant with the land. The only villages are in museum displays, and the grass shack exists only in old songs and stories. Lanai, the island where I lived, was being developed, and the cliff where I had camped under a pristine sky and watched dolphins play and whales nurse their young had been proposed as the site for a fancy restaurant. It was time to move on. I took my boat back to Lanai from Honolulu, studied maps of the South Pacific, and thought about islands. Some time later, deciding it was time to make a move, I quit an easy, secure job as harbormaster of a small state harbor. When I received Homer's postcard from Fiji, I booked passage on a six-week roundtrip charter flight that was to leave two days later.

Outside the small scratched window the moon flattens, turns burnt orange, and disappears into the sea. The big plane begins its descent into the darkness. In ten minutes we will land at the international airport at Nadi on the main island of Viti Levu. I haven't the slightest idea where I am going, or even where I am going to spend the night.

* * *

I cleared customs at 3 A.M. that morning and saw a big sign behind the immigration gates. The sign listed hotels and motels in the area and their rates. Most were prohibitively expensive. I chose a cheap one—The Sandalwood Inn, because I have always loved the scent of sandalwood—hired a taxi, and rode along a deserted road in the dark. When we reached the inn, there was a room available, plain and adequate. I fell asleep in the narrow bed and woke a few hours later, jet-lagged and disoriented.

I had crossed the dateline, gained a day and two hours, and entered the foggy confusion of fast-changing time and place. My eyes slowly focused on the glass louvers of unfamiliar windows. Through them the brilliant tropical sunshine shimmered on the crinkled red leaves of an immense croton bush. I lay quietly, listening to the splintering early-morning sounds of birds. I was in Fiji.

After breakfast at the inn, I wandered around wondering how to begin. The tourist brochures offered no clue, for I knew if I went where they beckoned, I would find tourists, and that was not what I was looking for. Someone told me I could take a local bus to Nadi, a town nearby, and I walked the half block to the main road and waited. The bus came on the "wrong" side of the street, in the English custom, and I boarded and set off. The bus was open and exotic, reminding me of buses in Mexico. Most of the passengers were East Indians, grave and unsmiling, the women elaborately dressed in nylon saris, gold earrings and bracelets, daubs of red paint on their foreheads and along the part in their gleaming orderly black hair. The Indians had a tenseness about them that did not draw me. The few Fijians on the bus were more appealing, bigger, more slack and relaxed in posture; the men flashed great white smiles at one another, and the women were quietly beautiful. I rode the bus to Nadi and got off at the marketplace.

Walking through the marketplace, I was again reminded of

Mexico. Here was food, real and unhidden by packaging: bright red piles of chilies; mounds of husked coconuts; bunches of taro roots; great sacks of rice; cabbages and beans, green and flourishing. The market was hot and humid, dimly lit with a soft blue light coming through makeshift plastic tarps, busy and confusing with the sounds of foreign speech. Walking among the stalls, I saw what looked like *yaqona*, great bundles of long dried roots. At one stall a group of Fijian men were sitting around a plastic basin filled with a brown murky liquid. I stopped and asked if it was *yaqona*. A man who spoke English said it was and invited me to join them in a bowl. I sat there, incongruous, a white woman among a group of smiling quiet dark men, and drank my first bowlful of *qona*.

It tasted faintly bitter and of the earth. After a few bowlfuls I settled in, relaxed, and was content to let the slow hum of the market float around me. Pita, the man who spoke English, told me about the *yaqona*, or *qona*. Grown from cuttings in the cooler wet places, it becomes a head-high bush with succulent heart-shaped leaves. The longer it is left to grow, the longer and thicker the roots become, and the stronger the effect. When uprooted, it is carefully washed and left to dry in the sun. Really clean roots produce a sweet-tasting *yaqona*; the bitter taste comes from the residue of soil on the roots. When it is fully dried, the roots are pounded into coarse powder in a log mortar, the *ta bili*. Then it is wrapped in a close-woven cloth and infused with cool water into a large, shallow, four-legged wooden bowl, the traditional *tanoa*. Served from a common coconut-shell cup, the *bilo*, it is the pervasive drink of Fiji, accompanying all ceremonial and social occasions. Visiting heads of state are greeted with a *yaqona* ceremony; even the Queen of England has tasted *yaqona*. The complexity of the ceremony varies with the importance of the occasion and the status of the participants.

The men at this particular stall were selling the *qona* root, and the plastic basin offered a sample of their wares. Men would come and go, put fifty cents on the worn plank counter, down a few *bilo*, and move on. A kilo of dried roots, or *waka*, depending on the season, cost from eight to fourteen dollars.

The more I drank the calmer and more comfortable I became. Getting drunk on *yaqona* is relaxing and quieting. Pita told me that if I drank a lot, my head would spin, and eventually I would fall asleep. I decided to leave because I wasn't sure that I was ready to fall asleep with a group of strange Fijians in the market on my first day in the country. As I left, Pita invited me to come back that night and drink *yaqona* with him and some of the employees of the hotel where he worked.

Feeling a little self-conscious, I came back at seven o'clock, carrying my small ukulele, to keep my date with Pita. We sat in the garden of the Nadi Hotel, slapping at mosquitoes, drinking *bilo* after *bilo*, and telling stories: mine of where I had come from; his of the Fijian way—*vaka Viti*, the life way, the way of sharing, and manners.

Three nights later, at a *yaqona*-drinking session at the same hotel, I met a young English-speaking girl, Desele, who was a distant niece of Pita's. Desele's manner was frank and curious. She came right to the point. Did I have a lot of money? A little. Was I married? No, not any longer. Did I have any children? Yes, one grown son who was a boat captain and who traveled a lot. Did I want to marry a Fijian?

That question stopped me for a few minutes, and then I thought back to how charmed I had been by Sam Whippy. I had been mate-seeking for years, ever since my marriage broke up, when my son was young. Western men were often eager for bed, but not often eager for marriage. "Maybe," I answered. "Why not? If I met the right Fijian, I suppose I might marry."

We talked some more, and I told Desele that I wanted to visit a village, a fishing village, where I might learn more of the Fijian way, and where I could go fishing and eat fish; and I wanted to sleep in a grass house. She invited me to go with her to her home village on the island of Galoa, off the coast of Vanua Levu, the other big island, and to stay in a *bure*, a Fijian sleeping house, with her and her grandmother. We would sleep by the shore of the sea; and there in the village, or *koro*, I would find the traditional Fiji.

Two days later we were up at four A.M. and on our way on a long journey: a six A.M. bus ride to Ellington to catch the boat to Vanua Levu, a four-hour boat ride through calm waters inside the great sea reef, waters dotted with the glowing green of shallow reefs, to disembark on an isolated landing and board another bus for a two-hour ride along a muddy road to a place that was nowhere. When we got off the bus, all I could see was a small wooden store with a billiard table crammed onto a cement porch and some Fijians and Indians sitting on the porch, listless in the heat. This was, Desele proudly announced, Nakadrudru.

I sat on the porch feeling isolated and lost, swatting at the flies that were biting my legs and ankles, while Desele chattered happily in Fijian to her friends and relatives. She ignored me, and I wondered what I was doing there and why I had come. After about three hours, she told me that her uncle had a boat we could take to her island, Galoa, and we walked down a small sticky path to a broad brown river. There, she bathed happily, and I, feeling apprehensive about dirty water, and a deep cut I had gotten two weeks earlier, bathed unhappily.

In the growing dark we set off down the river to the island, the small speedboat loaded with what to me were strangers. In a driving, opaque rainstorm that blotted out all signs of sea and horizon, we arrived on the island. It was velvet dark, and I stumbled into an old wooden house with the reassuring glow of a kerosene lantern and changed my soaking clothes behind a curtain.

From the other side of the curtain, I could hear the voices of Desele and her uncle Tevita. She was offering her *sevusevu*, a presentation of *yaqona*, to her family on her return to the village. The words rang ritualistic and strange, accompanied by resonant, slow handclapping that stirred me. I felt I had arrived in the traditional Fiji. When the *sevusevu* was finished, we went into a small thatched house nearby.

There was a bright fire burning in a corner hearth and a long low table set with a cloth, tin plates and spoons, dishes of giant clams cooked in coconut milk, and boiled breadfruit and fish. While we ate, the room filled with relatives: giant aunts who chattered about me, nodded appreciatively, and smiled in my direction. My friend was telling them that I had come to Fiji to find a husband, and they were suggesting possible candidates. I was intensely embarrassed. When we had finished eating, we went into another wooden house nearby, bigger, with corrugated-iron roof and wooden floor, to drink the *yaqona* offered at the *sevusevu*. There for the first time I saw the *tanoa*, the great wooden bowl, and paid attention to the speechmaking and the repetitive, cupped handclapping—three times when the bowl is passed, a single clap to receive it, and three claps when it is finished. Desele's uncle Tevita acted as my interpreter and guide, answering my questions and explaining customs. I was asked to play a Hawaiian song on the ukulele, the song drowned out by a continuous flow of *vinaka, vinaka* (thank you, thank you). Everyone stared at me and talked about me in Fijian. I understood only two or three words, but felt a warmth and friendliness I had never before felt as a stranger. After a long painful time sitting cross-legged on the floor, the *yaqona* was finished, and we went back to the first house to sleep on the woven floor mat, our heads toward the glowing kerosene lantern, a star shape of humans at rest in the night.

I opened my eyes. My companions of the night had vanished. Rays of sunlight gleamed through the old boards, patterned the mats. The room was big and bare. Two beds with mosquito-curtain frameworks stood at one side, shielded by a faded, flowered curtain, the one I had taken refuge behind the night before. A round table was decorated with a glass jar full of flowers and three speckled cowry shells. There was an old bureau with a looking glass and little more in the room. I could hear voices outside, calling, laughing, scolding. Feeling lazy and groggy from last night's *yaqona* session, I lay and stared at the still-glowing kerosene lanterns.

Desele came in, bright and talkative, "Let's go and drink tea with my grandmother."

After tea we explored the village. A tiny settlement of about fifty sleeping and eating houses, it was strung along the lee side of the island along a beige sand beach. Many of the *bure* had decorative borders of succulent ginger plants bearing heavy heads of fragrant blossoms. Some were adorned with giant clamshells facing the stone-and-sand foundations. Large dark green breadfruit trees shaded the paths. The village was neat and intense, mostly thatched houses, with a few wooden ones, clustered closely together in family, and extended-family, compounds. In the center was an old weathered wooden church, and behind that an unfinished concrete-block church of more pretentious proportions. A big cement water tank behind the new church caught the rainwater runoff from the corrugated iron roof of the church. The tank and a distant well were the only fresh-water sources for the village, and were the visiting and socializing centers of the women and children.

As we ambled from *bure* to *bure*, people called out from open doorways inviting us to come in and drink tea; shy smiling children followed in our wake. I was a star attraction, object of speculation and desire, for in the opinion of village Fijians, all Europeans, as they courteously call whites, are rich. The news that I had come to Fiji to find a husband, a slight exaggeration of my intentions by Desele, brought out offers and candidates. We went to visit cousin Ratu, sick with an earache, who was proposed as a husband. Ratu seemed disinterested in immediate marriage, so we passed on.

We crossed to the other side of the *koro*, a distance of less than two hundred yards, to the beach, where Desele said there would be a good place to go swimming, or bathing, as she called it. The water was warm and murky, but I splashed around for a few minutes trying to be a good guest, feeling self-conscious about my *sulu*, a length of cloth wrapped around my waist, which floated immodestly up around me. Anything approaching nakedness in women is *tabu* in Fiji, and I was wrestling with my *sulu* when there was a great splash next to me. A truly large woman with a dazzling smile surfaced a few feet away. She was Aunty Amele, delighted to be in on the swim. We communicated for a few minutes with smiles, and then Desele took me off to change into dry clothing. Aunty Amele insisted that we come back right away.

When we came back, there were a few women sitting in the sand under a big banyan tree weaving mats. Desele passed out the information that I was from Hawaii and was in Fiji to find a husband. I ignored her, trying to retain my dignity, and offered to show the women how to make a Hawaiian-style flower lei. Soon a great heap of frangipani blossoms was in my lap, proffered by laughing children, and I was weaving. When I finished the lei, they asked me to dance a hula; and I said I would like to, but there was no music. A man came out of a house nearby with a guitar and played a Fijian song. I danced awkwardly on the sand, everyone smiling and saying *vinaka, vinaka* as I danced. Then they all gathered on the trunk of the great banyan, and I took their picture. The man in the center of the picture, holding the guitar, was Malé.

Three days later I was sitting on a big smooth stone in front of Tevita's house watching the graceful sauntering back and forth of the village women, when Malé approached. He sat on a stone next to me, said, "*Bula*" (a greeting of life and health), and settled into silence.

"*Bula*," I answered, "do you speak English?"

"Little bit," and he returned to silence.

"You the one played the guitar for me the other day?"

"Uhmm . . . ," a low soft assent, and Malé lifted his eyebrows in affirmation.

I looked sideways at him, a handsome man with oblong dark-ringed eyes, luxuriant hair, a broad nose, and a moustache that molded full lips. His brown body, clothed in old rugby shorts and a faded sleeveless T-shirt, looked sleek and healthy. Malé's silence disturbed me, made me self-conscious, and I searched for something to talk about.

"Ah, you can tune my ukulele?" I asked.

Another murmur, and again that widening of his big eyes and subtle lifting of his eyebrows.

"Wait, I go get the ukulele," and I went into the cool darkness of the house to search for my uke. "Here, please tune for me." I found myself sliding into pidgin, a language that was often more comfortable than English after all of my years in Hawaii.

Malé reached out his hand. "Yes, give here." The knuckles

of his long tapering fingers were shaded with black homemade tattoos. His grammar was wrong, but Malé's accent was pure and soft, reminiscent of an English university. It fascinated me to be sitting on the stone foundation of a grass house talking with a muscular, barefoot Fijian whose voice sounded like that of an educated English poet.

Malé tuned the ukulele quickly, then strummed it in an intricate rhythm that showed he really knew how to play. After admiring the uke he said, "You come drink *yaqona* with me tonight, only us. I like tell to you something. You come?"

"Okay," I agreed, "what time you like I come?"

"When dark, come. Desele can show you." Malé slowly rose, handed me the ukulele, and drifted away down the sand path into the soft sunlit blur.

That afternoon I told Desele that Malé had invited me to come and drink *yaqona* with him, and Desele said, "You stay away from Malé; he's no good. You wait. I find you a good man." Beyond that she did not elaborate, but at dinner her uncle Tevita forbade me to go to the other side of the village.

"Those people are no good; they say bad things about our family. Everybody in the village knows that they are bad people."

I had the uneasy feeling that I was becoming some kind of property, valued because I had money and could help Desele's family with food and gifts. People I hardly knew were already telling me where I could go and with whom I could visit. Because I was their guest, I didn't go to drink *yaqona* with Malé, but I was ashamed, for I had accepted his invitation and felt that I had betrayed my promise. In less than a week social life in the *koro* had turned quite complex.

The next night Malé's sisters—Una, Ima, and Suli—came to talk with me on the beach. They told me that Malé wanted to marry me because his father had told him to. I explained that it wasn't the way of Americans; we married for other reasons and lived in an entirely different way as man and woman. The girls took all of that in stride, assuring me that Malé wanted to marry me because he was really good and always obeyed his father, and his father wanted it. I told the girls there was no way I could even think of marrying Malé until I got to know him and had spent time with him, until we had done things together.

"Good," they said, "tomorrow you come, and Malé will take you fishing. He is a very good fisherman."

It all seemed so strange and yet so forthright. My offhand remark to Desele had already brought me an offer of marriage. No one in the last twenty years had offered to marry me. I was intrigued. In the morning I told Desele that I wanted to find Malé that night and drink *yaqona* with him.

In response to my urging, Desele reluctantly took me across the *koro* in the dark, through an indecipherable maze of look-alike grass houses, to an open doorway from which drifted soft harmonious music. Peering inside, I saw a group of four men sitting in a circle facing one another. Fragrant white frangipani blossoms were tucked behind their ears, and a pile of blossoms lay on the mat between them. One of the men was Malé.

The music was lovely. The men sang with serious attention to one another in high pure voices; the *yaqona* in the *tanoa* glistened. Desele and I sat together on the mat. The men continued their singing as a number of other men drifted in, laughing and talking. Malé didn't look at me for a long time; then during a break in the singing, he handed me a blossom and said hello. He was quiet and serious, more involved with the music than with me. His voice was beautiful.

We stayed until the grog and the music were finished. Desele joked with the men; almost all of them were related to her in some way. I sat silently, watching Malé, feeling peaceful. He gave no sign that he was particularly interested in me, did not come to sit by me or speak. It was a man's party. It was almost as if I were not there.

Malé's indifference attracted me. Here was a man who was said to want to marry me, was spoken of as a very bad man. Nothing in his manner indicated either. But the singing captivated me. I love music and have always been drawn to those who create it. Malé was good, very good, and his seriousness and the purity of his voice seemed to speak for inner riches.

Tevita said nothing more about Malé. Desele told me that I should wait until her cousin Tovi came back to the *koro*. Tovi was a good worker, would make a good husband. It seemed she was trying to keep me as a prize in her immediate family. Malé was already married, Desele told me. He had a child somewhere, and many girlfriends. He was very bad, had been in prison for fighting. Her list was extravagant, smelled of jealousy. She stayed glued by my side, would not let me be alone with Malé.

Gradually Malé began to pay me little attentions: climbing a palm to offer me a coconut to drink, inviting me to the bush to gather mangoes, stopping to sit beside me in the shade and watch the slow *koro* parade. He was quiet and composed, ashamed to use the little English he knew, afraid he would say something wrong. I watched him from a distance: shouting and joking with the men, playing games with the young children. Fast and agile, he moved like a gymnast, like a dancer. Malé stood out, seemed different from the others.

Tevita had divided the village into "this side" and "that side." I should stay on this side and not go near Malé's house. I ignored him, went visiting. Malé's father invited me for lunch, the principal meal of the day. We ate boiled fish and fish soup, fresh greens cooked in coconut cream, and breadfruit. Malé sat next to me, picking out the best pieces of fish and putting them on my plate. The family treated me royally, asking many questions. The girls, who all spoke English adequately, acted as interpreters. Malé and I had not yet even touched fingers. It was a strange courtship.

Malé bloomed in the presence of his family; he teased and played with the small children, cuddled the infants, joked outrageously with his sisters. Their squealing laughter indicated that something was very funny. They obviously adored and feared him; he was demanding and often harsh, ordering them to wait on him constantly. Malé's mother was painfully shy, sitting in the background of the strong presences of her husband and son.

After lunch the dishes and pots were piled in the corner, the tablecloth folded, and the family reposed into tea drinking and animated storytelling. An old battered guitar, missing two strings, was brought out, and Malé and his father sang together, joking and laughing when Malé's father couldn't remember the words.

When the tea and singing were finished, Malé and I went outside and sat on the sand in front of the eating house. "Malé, is it true that you want to marry me?"

"Uhmm . . ."

"Why?" I asked.

"I just want to, that's all."

"But you don't even know me. Why?"

Malé did not answer. He had no answer. He looked out across the slate blue sea to the distant mountains, lit a cigarette, and played with his dog Sonny, who had come to sit next to us.

"Malé, Desele tells me that you are a very bad man, that you have done very bad things. Is it true?"

"I was very bad, very bad. I have done very bad things. But now I want to be good, to change. I want to follow God's way; the other way is no good."

"What did you do?"

"Fighting. I am very dangerous; everyone in the *koro* is afraid of me. I'm different from everybody. If I want to do, I do. I don't care. In Suva I walk around, drink, fight. I go to prison. The prison is very bad. They put me in a little dark room, no light, no bed, just a hole in the door to put the tea through. No food, just one cup of tea and one piece of bread. I stay there one week; then I come out. I stay in prison three months. I don't want to go back there; it's no good. I want to change my way."

Malé turned toward me; his thick lashes made his eyes look like they were ringed with kohl. "Joana, you can help me change; the European way is good, soft. Maybe you can help me."

He stood up. "Come, we go. I go help my father in the *teitei* (garden). You come tonight to Qare house. We play music, drink *yaqona*. You come."

Qare's house was nicknamed "The Blue Corner" by Malé's sisters and cousins to distinguish it from the red corner, which was Malé's house and off-limits. I told Desele I wanted to go there and listen to music. I wanted to go alone, but she insisted on coming along. When we arrived, the small *bure* was crowded. Desele acted very aggressively toward Malé, as if she were enamored of him, flirting and teasing. Malé sat next to me, passing the *yaqona*, lighting my cigarettes. "Joana, dance a hula for us; we like to see you dance."

It was not easy to dance a hula without knowing the words, but I did the best I could. Desele got up and danced with me, everyone shouted and laughed. It was fun. Little by little people drifted off until only Desele, Malé, Desele's cousin Veresa, and

I remained. Desele insisted we leave, "Come, come, we go home now." I didn't want to leave. I wanted to stay until everyone had left, to be alone with Malé. I wanted to sleep with him. To know him.

Desele would not leave. I lay on the mat next to Malé, slack and groggy. Desele and Veresa got into the bed amid much joking and teasing between them and Malé. After a while of tussling, Desele came and lay on the mat on the other side of Malé, and Veresa lay down next to her. We all slept.

In the middle of the night, I felt Malé's body next to mine. "I want you," he whispered.

"Why?"

"I just want. Do you want too?"

"Yes," I answered, "I want too."

In the morning Desele and I went back to her house. I was sitting on the mat drinking tea when a man I did not know came in and said he wanted to talk with me. He was one of Desele's uncles and spoke English. He explained that Desele's grandmother had asked him to come and explain Fijian custom to me. If I wanted to see Malé, he must come here. I was not to go to him. If I wished to stay with Malé, I should leave the *koro* and then come back and stay with Malé's family. I said I understood, was leaving soon, and would abide by their wishes. I thanked him for coming to talk with me.

It was clear that if I wanted to know anything about Malé, I had to be alone with him, and that was impossible in the *koro*. My plane was leaving in a week, and I asked Malé if he would like to come and spend a week with me in Nadi before I left. If I could get him a visa, maybe he could come to Hawaii for a little while. It would give us some time together to see if we had any basis for a relationship. Malé answered immediately, without question. Yes, he would come.

The night before we were to leave, Malé's father called him to come to talk with him. I sat in the dim golden light next to Malé and listened to his father's voice; strong and serious, it sounded almost as if he were praying. He talked for a long time while Malé listened silently. When Malé's father was finished, we went outside.

The night was warm, and stars glowed in a humid sky. The world felt watery, calm. "Malé, what did your father say just now? What did he want?"

"My father tell me he is old now and can't work like he used to. He needs me to help him, to fish and help in the garden. He says he doesn't want me to go away with you to Hawaii. He says if I go away for a long time, he wants a boat and outboard engine and new fishing net. That way he can take care of the family if I go. He wants me to come right back, just go to Nadi and then come home. He doesn't want me to go around in Suva, just come right back."

"Okay, Malé, we just go to Nadi, and then you come home. You don't have to come to Hawaii with me."

Malé took my hand, stood, and pulled me up alongside of him. "Come, we go walk around the *koro*." Late that night, as Malé was walking me home to Desele's house, we stood in the shadows of a breadfruit tree.

"Joana, when we go to Nadi, there is something I want."

"What, Malé, what do you want?"

"I don't know whether you can or no."

"What, Malé, what is it?"

"I want some shoes."

"Sure." I smiled in the darkness. "I can get you some shoes."

"And there is something else I want."

"What?"

"Some trousers and a nice shirt."

"Okay, Malé, I'll get you some shoes and trousers and a nice shirt."

"And there is one other thing I really want."

"What is that?"

"I can't tell you now. I tell you in Nadi."

* * *

Nadi was hot and sticky. We stayed at the Nadi Hotel and spent most of our time sleeping and eating and taking showers. Malé shopped intently and with great care for his clothes. He chose a good pair of shoes that looked as if they would last, and he cleaned them thoroughly when they got the least bit dirty. He knew exactly what he wanted: a certain shade of light blue for a shirt, a particular material, blue, for his trousers. The new clothes transformed his ragged village look.

A watch, he explained, was the other thing. I asked why he needed a watch in the village. Was it just to show off? No, it was so that when he was fishing he would know when the tide was coming up, when to check the net. It sounded reasonable, so I got him a watch—not the shiny gold one he coveted, but an underwater watch that would withstand work and water.

It was the first time in his life Malé had stayed at a hotel. He was shy and delighted by everything, couldn't wait to go home and tell the story. He posed for innumerable snapshots: in the swimming pool showing off his new watch, in front of the hotel in his new clothes, in the room lying in bed and playing the ukulele. I promised to send the pictures to him from Hawaii.

There was not much we could talk about. I had learned to say hello, goodby, yes, no, and thank you in Fijian. The rest was beyond me. I bought a Fijian dictionary, and we struggled with it; but it was in the Bauan dialect, and Malé spoke Galoan, the dialect of his village.

Slowly we sorted through the intricacies of our negotiations. Malé wanted to marry me; he would not go further than that. If I came back, we could go live together on his family island, Vedrala. The family had its own private island, a nice one, close to the village. His father would give us the island and build us a *bure* there. If I came back, he wanted a new boat and an outboard. He wanted me to build his father a small wooden house. His father was getting old, and it was hard for him to keep rebuilding the family's grass houses. The whole discussion was archaic and honest. I knew that Malé did not love me. He was following his father's wishes. But he was kind and attentive, and seemed solid inside.

One night we went dancing, and he got very drunk. A boy came over and tried to take one of our beers off the table. Malé stood up and, his eyes clouded, became opaque and menacing. He was lost somewhere in his own private world of rage. He wanted to fight.

"Malé, Malé." I took his shoulder and looked into his eyes hoping to draw his attention away from his anger. There was nothing inside his eyes. The Malé I knew was gone.

A cousin who was with us said, "Don't worry, Joana. It's all right. Just leave him alone."

Malé stood for several minutes, fists clenched, body tense with fury; then the boy backed down, put the beer on the table, and apologized. Malé stood for several minutes, then said, "You wait here, Joana," and disappeared into the gloom of the dance floor. When he came back, whatever had possessed him had passed.

Lying in bed that night, I looked at the reflections on the

ceiling from the hotel lights outside. Malé had just fallen asleep, and I heard him moaning, a muffled cry coming from locked teeth. I turned to see him lying with clenched fists, rigid, as if in bonds. I thought he was dreaming and let it pass, but the moaning continued. As I was about to wake him, he shouted something in Fijian, turned his head toward me, said, "Joana," and opened his eyes.

"What was that, Malé? Did you have a bad dream?"

"It was the *tēvoro*. He was outside the door trying to come inside. I cursed him, and he went away. When the *tēvoro* comes, you cannot move. The only thing you can do is curse him; then he go away. No, it was not a dream; it was the *tēvoro*."

In the morning I lay in bed thinking about me and Malé, and Desele's warning, and wondered if he had indeed changed his ways. When he woke, I asked him about the *tēvoro*. He couldn't explain, so I went to the dictionary and found: "*tēvoro*, an evil spirit, a demon. *Sa vuka na tēvoro, ka sa tū na tamata*. Proverb of a man who after evil living comes to his senses. Literally, the devil has flown, but the man remains." It seemed an omen, as if the dictionary had answered my question.

We went to the airport about nine that evening. My plane was to leave at eleven. We spent two uncomfortable hours waiting, said little. Malé said he wanted me to come back and marry him, we could go to Vedrala, it was really nice there.

"It would be good," Malé told me. "Be sure to come back."

I told him I would go home and think about it and come back in April, four months later, and then we would see. At the boarding call I left Malé in front of the locked doors of the international waiting room; he could go no farther.

He did not kiss or hold me. Just as the doors closed, he waved and said, "Come back, Joana, please come back."

The plane was late taking off, and I sat in the confines of the cabin wanting to bolt back to the airport. From where I sat I could see Malé, a dim distant figure, holding onto the chain-link fence surrounding the runway. The plane taxied to the far end and then revved up for takeoff. As we passed the fence and lifted off the ground, I could still see Malé standing there.

BACK HOME

Honolulu seemed garish and sharp after the soft light and laughter of the village. I checked into an airport hotel to spend the night until I could get a connecting flight back to Lanai. Sitting in the hotel dining room, I looked around at the shiny mirrors, the glittery reflections. Knives and forks scraped on porcelain plates, and voices were strident with drunken laughter. I was only eight hours from Malé, and already missing him. I was alone again, surrounded by luxuries, and dull and empty inside.

When I returned to Lanai, my house was quiet and cold. Only my small dog, Jackie, was there to welcome me. When Malé and I had left the village, a crowd had gathered on the beach to wave goodby and to urge me to return. Now I was home, facing each day, wondering where to go and how to fill the hours. I had left my job because I felt that I had to open up my life, to make room for something new to enter. There was nothing new on Lanai, only the same places, the same faces.

I wanted to talk about Malé, to share my adventures. I called my mother in Los Angeles. I told her of Fiji, of my experiences in the *koro*, that I had met a young, handsome man with his own private island. He wanted to marry me; I might go back. My mother was furious, thought I was crazy, that I should be committed. I understood her rage. I am her only child, and she wanted me close to her in her old age. I felt sad that she could not share my excitement.

My mother and I have not lived together for years, ever since I went away to college. My father died when I was six, and my mother struggled to raise me on her own.

I grew up alone, have dim memories of faceless babysitters. Books became my companions. I would check out eight or ten from the library and read them one after the other, sitting in the sun porch of a house in Chicago where my mother rented a room. I dreamed of faraway places, mourned for fictional suffering animals, yearned for affection. My world was private, moody, introspective. I loved my mother, and she loved me, but she was rarely home. When I grew older, we often fought over my refusal to want the things she wanted for me: a respectable professional husband, a nice house, the comforts and security of money. My mother had done the best she could, but we did not share the same values. I certainly could not expect her to understand me now, but her anger made me feel depressed and lonely. She and my son, Ian, were the only family I had.

I called Ian, who was working as captain of a charter yacht in Santa Cruz, to share the news. Ian and I are close, have always been close, even though his work and life have often taken him far away. Although we rarely see each other, we share a common love of the sea and a need for the unconventional. Ian thought my story was great. I could visualize his grin on the other end of the phone.

"Sure, Mom, if you think it could make you happy, you should go for it. You can always come home if it's not right." The idea of inheriting a big family in Fiji charmed him. Ian's approval was reassuring. I knew that he would support me in whatever I wanted to do.

I tried to get back into my routines: practice hula, clean house, drive down to the beach and swim. I hiked in the forest, picked flowers to make leis, visited friends, went to the small hotel at night to drink with the boys. The days and nights dragged by.

I was lonely. I had met a man who wanted to marry me. The reasons were all wrong in terms of my own culture, but the offer was real. Life in the village was hard, physically uncomfortable. There were none of the things Americans valued. Yet standing in the luxury of my shower, I recalled bathing from a bucket under the stars. The steak and potatoes and salad I ate alone in my kitchen tasted flat compared to the boiled fish and breadfruit I had eaten surrounded by the lively conversation of Malé's family. I bought good wine to ease the loneliness, but it was no substitute for the grog parties; and the music of my radio could not replace the seductive warmth of Malé's voice.

I looked in the mirror; wild bushy sun-bleached hair, glasses, small eyes, self-conscious smile. I was no beauty, and it was unlikely that anyone would want me for my looks; at least no one had. My body was good, strong and healthy; the sexual desire that had driven me into believing that I was in love with countless men was starting to fade. I could no longer confuse lust with love. Love, I was beginning to believe, was something that was constructed, created by attention and thoughtfulness. I knew Malé did not love me; yet perhaps he could learn to. He wanted my help because he loved his family. That was to his credit. I did not yet love Malé, but was attracted to him and to the life he offered. There was no question but that it would be hard, very hard. Our cultures were separated by more than thousands of miles of blue sea.

When I was an undergraduate, I wanted to be an anthropologist, studied at UCLA and Berkeley. Primitive cultures fascinated me; I felt drawn to peoples who had retained mastery over their immediate worlds. I respected Malé's accomplishments, his strength and agility, his skill in fishing, his knowledge of the natural environment and how to use it. If I wanted to eat, he could catch a fish for me, climb a coconut palm, plant a garden, build a fire. His skill with the machete awed me; his love for animals, unusual for a Fijian, comforted me. He could not be all that bad if what I saw of him seemed good.

I had told Malé that I would come back in April because I felt I needed time to reflect, to consider, and needed the time to make certain that his offer was, at least in his own terms, substantial. Yet as I looked toward the next four months, I felt I would only be passing time until I could return and find out more about Malé and the village. There was also the danger that old habits would settle in and the dream would become more and more impossible and remote. Perhaps I would never return. If I wanted to explore the reality of his fantastic offer, there was not much sense in waiting.

Down at the harbor one night I ran into my old friend Françoise, who had led a lonely sailor's life for years, rejecting the offers of countless men who wanted to bed her. She was extremely beautiful and competent and was waiting for the right man. She had finally found him, a bush helicopter pilot named Steve, who was as competent as she. They were going to get married. Françoise called me over to meet Steve, and the three of us sat in the tiny cockpit of her boat and shared some wine and talk. I listened to their story then told mine, leaving out none of the details.

"Do it, Joana, do it." Françoise told me. "It sounds right; you should just go back. Don't wait." Late that night I left the two of them and went to sleep on my boat. Lying in my bunk, listening to the electric crackle of the sand shrimps beneath the hull, I decided to go back to Fiji.

⯭ ⯭ ⯭

The connecting flight on the small plane from Nadi to Labasa was scary. We flew low over the rugged mountain spine of Vanua Levu, and I looked down on the spiky peaks of the mountains with some fear. The pilot was busy chatting with his copilot, a very young Indian boy who was learning to fly. We bounced hard, and I wished the pilot would fly and leave the teaching for a better time. Yes, I was back in Fiji, the land with lots of laughter and talk, and little attention to details.

At the market in Labasa, I searched for a taxi to take me to Nakadrudru and ran into one of Malé's aunts. She greeted me warmly and managed to convey the information that Malé was in Suva. My heart sank. Had I come back all this way to find out that Malé had not returned to the *koro* as he had promised? Were Desele's stories of wives and girlfriends all too true?

During the hot dusty two-hour taxi ride to Nakadrudru I wondered if I was, as my mother had so confidently pronounced, crazy. My suitcase was loaded with gifts for the family—T-shirts from Hawaii—and enough clothes to last many months. I had also brought my other ukulele to add to the potential of Malé's quartet. I was nervous and apprehensive, perhaps it was a deception after all.

When we arrived in Nakadrudru, Una, Malé's sister, was sitting on the porch of the store.

"Joana," she cried out in glee, running toward me. "Joana, you came back," enveloping me in her substantial arms. "Oh, Joana, how good you came back." I was embarrassed to ask about Malé, chatted with Una about my trip, did not say anything about why I had come back.

In the late afternoon a boat was going to the island, and we reached the village in the splendor of the setting sun. I looked anxiously at the beach in front of Malé's house for a glimpse of him. His father and mother were there, his sisters and brother, but no Malé. News that I had returned had already reached the *koro*; how, I don't know, but the family was standing on the beach waiting for the boat. Malé's father greeted me with tears in his eyes, kissed me in the Fijian fashion, that is, sniffing the side of my cheek. His face was bristly and warm. But still no Malé.

"Malé here?" I asked. Ima told me that Malé had gone fishing with his younger sister Boko. They had paddled to a distant reef, would fish all night, returning in the morning. No, he had not gone to Suva; he had come right home.

Dinner was waiting. The family had hustled to prepare a feast for my return. No one seemed perturbed that even though I had said I would return in April, I had come back in two weeks. In fact, everyone seemed delighted. Malé's father indicated the good bed in the eating house. "You, Malé sleep here." It was all so easy. In their eyes Malé and I were already married. The only thing missing was Malé's reaction.

After dinner and tea Malé's father asked Una if I wanted to stay the night or go find Malé. I told her I would like to go if there was a boat we could use. That was no problem, she said; we could borrow a boat. I had brought some *yaqona* for my *sevusevu*, and Malé's brother, Vili, went off to pound the *yaqona*. We loaded the borrowed boat with the *tanoa*, the pounded *qona*, some mats and pillows and set off in the deep rich dark of a moonless night.

It was caressingly warm in the mangroves as Vili poled the

boat through the maze of the channel. We had no light, and the stars' reflections glimmered on the calm water inside the reef. The sky was brilliant; there was enough light from the millions of stars to find our way. "Una, where's Malé?"

"Over there, on the other side." We came out of the mangrove channel to the far side of the reef. I could see no light, no sign of Malé and the boat. "He's in there," said Una, "that side. Just call. Call Malé."

"Whooee, whooee," I called into the darkness. "Malé, Malé." From inside the mangroves I heard Malé's voice, speaking Fijian.

"Call again," urged Una.

"Whooee, Malé." Then I asked Una, "What's he saying?"

She was enjoying the joke and laughed, "He wants to know who it is. He doesn't know it's us."

"Malé," I called into the darkness.

From inside the dark foliage a dim light slid toward us. As the light grew closer, I could see Malé standing on the bow of the boat, poling it toward us.

"Malé," I called again.

The boat came close, and Malé shaded his eyes from the glare of the benzine lamp. "Joana?" he called, "Joana?"

"*Bula*, Malé, I came back." I stood up on the bow of our boat as the bow of Malé's boat touched ours. Malé reached out his hand, warm and strong, and took mine. "*Bula*, Joana, *bula*."

Vili and Una exploded in laughter; Boko giggled. We all sat in the borrowed boat and drank *yaqona*. Malé, Vili, and Una chatted in Fijian; I could understand nothing. After a while I went and sat on the bow of the boat. Malé came and sat next to me. I wanted a hug, a kiss, some sign that I had done the right thing.

Malé touched me lightly on the shoulder. "It's good, Joana, good you came back."

DREAM ISLAND

Today is going to be one of those fierce hot Sundays. The sun is a faint pink smear on the horizon, the sea mirror calm. A big pink-white luminous cloud floats motionless over the spur of the mainland. Thankfully the village is still sleeping. I look at the path reflected off the clouds onto the water. I wish my heart to be as still and calm as the sea, as motionless as the clouds.

The typewriter is new; these are the first words to come from it. I must choose them carefully, just as I choose my clothes or the cut of my hair or the set of my smile. Words, the ones inside that rage and fret, and the ones outside that wound and tear, that can give intense pleasure or pain, are the life blood of the mind. I love you. I do not love you. Words.

Malé's eyes grow dark and mirror a deep pool of darkness. I look into them and see a well of pain, of distance. Leave me alone, the child cries in the night, and the words mean: comfort me; take away this unexplained burden of fear and thought. I look into Malé's dense brown eyes, the archetypal eyes of darkness, and say, "Tell me, tell me what is wrong."

"No," he says flatly, far away.

No. I do not love you. No. I do not want you meddling in my heart, poking around in my sorrow. Stay out, stay away. Buy me presents and leave me free to enjoy them. Ask no price, extract no cost, just give me and go away so I might be happy and free from your need for gratitude.

The sun is higher now, and the cloud has gone. A small boy is watching me type, and Malé's teenage sister Boko has brought me a cup of coffee. She is dressed in her pink Sunday dress, an incongruous fluff of pink ruffles now squatting by the fire, blowing life into the morning tea. The tide is going out, but is still high, and our borrowed boat floats close in, pointing to the exploration of this dream. There is smoke coming from the family eating house, and the convivial pig is wagging her tail and wandering under the ginger blossoms in search of breakfast. She just tugged at a mat and, finding it unpalatable, moved on.

The typewriter is my rock and my guide. It will keep me from falling into hopelessness. The words, the images, must be chosen to remind me that it is inappropriate to search for love. Love is a gift that is given and taken away. I cannot buy Malé's love. I can only buy him things.

This past weekend we went to Labasa, where I bought Malé a spring: a steel contraption to strengthen and increase the muscles of the body. He wants to train, to become bigger and fiercer and more handsome. He says he is willing to give up smoking cigarettes to achieve that. I bought myself the typewriter, a contraption to clarify the journey, to make it clear to myself why I am here, and to keep distant the loneliness.

The sun is about to rise over the hill, the sky a marbled blue-gray. A faint breeze promises. Boko, Ima, and Una have just come to cluster around the typewriter, smiling and curious. They want to be part of this new toy, to explore these exotic possibilities. I ask them to leave me alone.

They are becoming used to my curious ways. At first, when frantic with the frustration of trying to understand why I had chosen to come to this strange and difficult place and give up my fortune here, my meager fortune, to buy the love of this fierce, charming man-child, I would search for a small spot of solitude to shed my tears or let my thoughts fall on the sand and sink into the earth. Then the girls would come chattering around me to sit by my side and make the small talk necessary when one cannot speak the language. They could never understand why I wanted to be alone, what it meant even, for there is no word, really, in their language for alone. Now I am learning a gentler way of driving them away than the tense emotional rebuff. Often I feel guilty when I see their kind intent and know that I have yet to wallow in my thoughts, now so much my way that to change it seems unthinkable.

This morning I look at the coil of rope hanging from the banyan tree that last night shone so menacingly in the glare of Malé's new flashlight. We were sitting on the overturned punt, and I was trying to coax him into talking to me, into explaining why he was so angry that he would not speak to me.

I knew why he would not talk, for I had said, "No, no more money, no more indulgences, for you lied to me about the money I gave you." And he took away his affection and turned to silence. The coiled rope looked like a man hanging in the shadows of the tree that is the house of the devil, the home of the *tēvoro*.

The *tēvoro* lives in the baka tree next to the sleeping house, and has long hair and a name something like Roko. He visits in the night, and when he comes you cannot move.

The *tēvoro* has visited often since that first night in Nadi. Sometimes Malé shouts and mutters in his sleep; he is fighting with the *tēvoro*. I try to remember the words, to ask about them in the morning, but the Fijian is a garble to me, and I remember nothing.

I am beginning to perceive Malé's complexity. His joking and laughing hold at bay some deep misery. I have no clue as to the cause. He is fine with me as long as I let him do what he wants, give him what he asks for. When I do not, he turns away and loses himself in the family. He is spoiled; his sisters wait on him constantly, and everyone seems afraid of him. But it is something other than just being spoiled. He says it's because he wants to be happy every day; if he isn't happy then he has "the trouble inside."

If I understood Fijian and could talk with him in his language, it would be easier, but Fijian is beyond me. This is no place for formal study; it's too hot, and everything is chaotic—no schedule, no order. The girls are happy to answer my questions, but soon lose patience and turn to chattering in their own language. There are no dictionaries or grammars in this dialect, so I can learn nothing from books. It's strange; I, who have been so articulate, am now so dumb.

⯭ ⯭ ⯭

I have a new audience. Vili, small Suli, young Jone, and Va are watching the mysterious intricacies of the typewriter as these words appear on the page. Suliana, the youngest, is most interested in the flash and the noise. The words themselves are of little value, but anything new and bright is worth her attention. The breeze floats the page over the keys; everyone is intent on this process. I am trying to free my mind from their watchfulness and return to the subject. It is impossible, for the subject is found in the dark cells of loneliness, not under the intent gaze of the family. I feel that I should continue typing, for this little spot of newness is the morning's fascination. Then my fingers stop over the keys: what can I say if the only reason for saying it is to provide the flash and clatter of the keys. The audience has now increased by three. There are five small children, three adults, and one young boy gathered around me, bent over the typewriter, helping me forget how serious is my mission. It is good that my cultural intentions are interrupted by this curiosity, for it reminds me that life is lived better as a joke than as a tragedy, and of what account is the sleeping, unspeaking Malé when I have an audience of seven watching the words flow onto the page. For that matter, what difference are the words when it is the novelty of the process that has so great a value. The novelty will soon wear off, however, and then everyone will drift away to some other diversion, and I will be here with the morning flies and my Sunday-morning mind.

It is good to begin the day with prayer. I have found some words that seem to wear well, to hold up under my inconstant belief. "The Lord is my shepherd, I shall not want . . ." It's the part about the green pastures and the still waters, the quietness, that comes when I remember the existence of cool calm places where the presence of God is undisturbed by human confusion.

Toka, Dimate, Maria, Vulase, Sione, Kavaia, and Suliana are all staring at the keys. I have just typed their names to shelter me from a chaos of sounds which have not yet separated into words of meaning.

"The Lord is my shepherd, I shall not want. He maketh me to lie down in green pastures, he leadeth me beside still waters, he restoreth my soul . . ."

It is Sunday and time to consider the love of God, and forget for a while the love of Malé still sleeping in a bed I had to leave before he would lie down in it.

Malé wakes from a dense hot sticky sleep and looks at me mutely, makes a low sound, questioningly, in his throat, and shifts his eyes ever so slightly. It is a question. Is it okay? Am I forgiven? I look the other way, yet my heart is already soft, for how can I hold a grudge against this beauty. He falls limply back into the bed, and my heart hardens again. I renew my purpose: to pay attention to the whole life around me and not get caught in the universe of those eyes.

I have been living with the family almost a month now, a month of slow discoveries. Just getting their names straight was hard. I can't make much sense of the sounds and often cannot tell the difference between names and other words. The head of the family is Jone, Malé's father, whom I call Momo—uncle. Malé's mother is Veima Saiyabo, Nei, or aunty, to me. Then there are sisters Suli (Suliana Mere), about age 30; Ima (Sereima Vakacequ), 20; Una (Unaisi Moceveicoi), 18; Va (Vaseva Moroci), 16; and Boko (Sisilia Boko), 14, the youngest. Malé's oldest brother, Sione, 32, lives away from home on Taveuni Island, as does his oldest sister, Luisa, 34, who lives in the capital city of Suva. Both Sione and Luisa are schoolteachers. Malé's other brother, Vili (Viliame Uwate), 27, lives here in the family compound with his wife, Banoko, 24, and their small children, Suli (Suliana lailai), 3, and Vili (Viliame lailai), 1. *Lailai* means small and is used to distinguish the small children who have been named after uncles and aunts. Sister Suliana's children, Sione, 8, and Malé, 2½, also live with us. They were sired by different fathers, both now only a dim memory.

It's crowded. The family's big *bure*, called Botoi—houses also have names—is dismantled, awaiting rebuilding. That puts us all in one eating house and Viliame's small *bure*.

Family names are confusing because of the custom of naming children after uncles and aunts. Everyone in the family is named after someone else. There is the lineage of Jone—Malé's father—who is named after his father; Momo's oldest son, Jone (Sione); Suliana's oldest son, Jone; and in Taveuni, Jone, the first son of the first son. Four living Jones in the family. To add to the confusion, Jone is sometimes called Sione, and in English John. Malé is named after his uncle Malé; and small Malé, Suliana's youngest son, is named after his uncle; and so it goes. The girls are all named after aunts or grandmothers.

The family is huge. Everywhere we go we meet uncles, aunts, and cousins. Malé's mother's and father's sisters and brothers are considered little parents. Malé calls his father's brothers *Ta* (father) and his father's sisters *Nei* (aunty). First cousins, called *tavale*, are thought of as brothers and sisters. All of the brothers and sisters of one's grandparents are also called grandmother and grandfather. Sometimes I am totally lost when Malé tells me a story.

When I came back before Christmas, Malé's father moved out of the one good big bed in the eating house and into Viliame's *bure*, and Malé and I were given the bed. Suli, Ima, Una, Boko, Va, small Malé, and small Jone also sometimes sleep in the eating house. People seem to care little where they sleep. Just find a pillow and lie down on the mat. The kerosene lamp burns all night, and often one of the girls is up in the middle of the night to boil the tea for Malé's father, or to cook some fish when the boys come back before dawn from fishing. There is no privacy. I dress behind a curtain, and we drape *sulu* over the mosquito net for a hushed and hurried intimacy.

Everyone treats me well, as one of the family, but favored, and I am not allowed to do anything for myself, not even make a cup of tea. I think they think of me as weak and unable. I have always prided myself on my strength, but here I am weak compared to the girls, who carry one hundred pounds with ease. I can do nothing for myself; the kitchen is so disordered that even if I had the strength in this heat to cook on an open fire, I wouldn't be able to find a clean pot or plate or spoon when I wanted to. Everything is beyond my competence: washing clothes, carrying firewood, cooking, fetching water. It's the first time in my adult life I have felt so useless. Malé's mother often scolds the girls when she thinks I need something. It's considerate, but humiliating.

Days and nights blend. There are no jobs, so no one has to be anywhere on time, but everyone cherishes a watch. Fishing is often done at night. It's too hot now to move around much in the daytime; at midday it's impossible. The nights are for drinking *yaqona*, making mats, and visiting. The very early mornings and late afternoons are for working in the garden, and the midday is for sleeping.

This is the hurricane season, which starts in November and lasts through March. Right now the humidity is almost 100 percent. A half mile away I see a heavy gray curtain of thick rain between us and the mainland of Vanua Levu. Very little moisture separates us from that rain curtain. It is moving this way, and until it reaches us the heat will be almost unbearable, fit only for sleeping.

Sundays are often like this, for there is no work done on Sundays, and the inactivity combines with the humidity to produce a sticky dream day of restless sleep and brushing away the flies. The village blooms for a few hours when everyone dresses up for church: clean formal *sulu* and fancy incongruous tops on the young girls. Then after an hour or so in the stifling church, listening to a sermon in a language which, to me, often sounds angry, everyone goes home and changes the bright clothes for everyday ones, and the flock of bright butterflies is gone for another week.

After lunch Malé comes to me and asks if I want to go with him to Vedrala. It is pouring rain, and the air is heavy and gray. I stifle a desire to say no, not now, for this is in the nature of our nuptial flight to the place of my dreams, the private uninhabited island where I am to make my life with Malé, the dream life; and this is the first time he has asked me to go over alone with him: a reward for tolerating his withdrawal last night. So at this improbable time, in the stinging rain, unable to see five feet ahead of us, we will go to Vedrala, to begin our exploration, to look around.

I remind myself that everything has its own beginnings, and

that to carry the past around is asking for a heavy burden, a house built of bitterness, so I agree. Then I have to smile at the wildness of this young savage. For all of my liberalness and desire for the primitive life, my tendency is still to want to control, to civilize and make proper and respectable, and there is no way that I will ever be able to control Malé. That, in addition to his beauty, is why I am beginning to love him.

The rain streams down, pelting the water. The boat goes fast, and the drops sting my eyes so that I cannot look ahead. I glance sideways at Malé; he's wearing only shorts, his thick woolly hair adorned with raindrops, shimmering diamonds in the dark, his hand over his eyes—mysterious, able. Who could not love his intense ability, his mercurial power? A few minutes later we slide through the mangroves that ring the island, and I look at the bright green leaves of the young trees. The boat slows; we are in the calm of the mangrove channel, and the journey enters the dream time.

The rain grants a peculiar beauty to the mangroves, cleans and polishes the twisted branches, washes the leaves and turns them into bright green promises. The rain cleans the water, clearing the mud; and as the boat glides through the channel, I realize how appropriate it is to begin this adventure in the cleansing blessing of a Sunday rainstorm. Malé is in good humor after his sulk last night. He is aglow and healthy, wet and glistening, full of joking vigor. He is very sexy; rain trails down his chest, along his muscular arms, wet woolly hair glistening, eyelashes dripping rain and salt. His smile is bright and happy. I look at him, and my heart melts. All is forgiven.

Vedrala has the heavy silent calm of an uninhabited place, a place where nature is untarnished by human design. This is what I have longed for: the Pacific island, the dark-eyed mate, the silence, the long nights and long days. Stepping over the threshold of my desire into reality, I see clearly the vast channel between them. Now I must live this life, not dream it.

We throw the stone anchor onshore and climb out of the boat into the cool fresh water. The sandy shore is littered with fallen coconuts, broken black branches, trunks of trees, fallen nuts and leaves. There are no footprints, no man or animal to disturb this silence. The rain has slowed; a soft coolness prevails. We walk over to where we will build our first *bure*, and Malé points out where we will have our houses; one on the hill for the wind to cool, one on the shore by the mangrove channel, and one in the vague distance.

Everything here is unknown. I do not know what bites, stings, pokes; what is poison; what is edible; what is to be used for what reason. This then is my education: to learn this place, to make it home. Malé will be my teacher. What I will teach him is uncertain; a gentler sense, a kindliness, I hope. Mostly I want to teach him that there is no need for that darkness to enter his eyes, the darkness that blots out thought and reason, that makes him want to smash and hit and flow with blood, to wash clean with anger.

Vedrala. I looked in the dictionary for a meaning. The only one I found was *dra*—blood.

Dream island: the soft rustle of rain in the leaves, the bright bloom of rain wet on the rust orange trunks of the coconuts, the leaning tangle of unknown vines, wet sand gritty underfoot, wasp nest swinging from the branch. In the wash of the mangrove channel, where the waves come in and clean the swale, are great piled nests of unknown insects: ants, termites, I do not know. But many, many of them in an immense labor, an unaccountable labor, grain by grain, have constructed these huge lumps, these sand piles designed by God. Malé pushes off the top of one of the hills. His motion is quick and careless, yet accurate. It is the way he moves, like a coiled spring springing, like a flag whipping in the wind, flinging himself into activity. Inside the nest is a perfectly round hole about four inches in diameter leading down. It is lined with smaller grains of sand, smooth, almost polished. There is no sign of anyone living in there, just the drops of rain making the grains move as if alive.

No one has lived on Vedrala for a long time. A small pile of rocks marks the foundation of Malé's uncle's eating house, used briefly in the early sixties, then abandoned. A large pile of rotting coconut husks is where someone, long ago, made *copra*.

I look into a landscape of small trees draped with heavy vines: large mangoes, silky ironwoods, hundreds of coconuts, and, above us, the shadow of great rain trees with feathery soft leaves. The brush seems almost impenetrable. Eight-foot-high grasses with razor-sharp leaves block our passage. The ground is chaotic with broken branches, leaf litter, old coconuts, vines. Malé follows an overgrown trail and leads me into the bush.

The island seems bigger than its forty acres because of the intricacy of the vegetation. I could travel here for days, around and around, apprehensive of the wasps, with the raised-hair feeling that comes in wild thick places. Malé strides in front of me, barefoot, swinging at vines and saplings with his machete. He stops and chops his name into a coconut trunk. He is at home.

We climb to the top of the small hill and look around. There, to the east, less than a mile away, is Galoa, green and lush in the gray light. A swath of mangroves is all I can see of Tavea, a mile or so distant; to the south and beyond Tavea is the broken-ridge outline of Vanua Levu. Seaward, beyond Galoa, some five miles distant, is the high-ridged outline of the big island of Yaqaga. Looking north, Malé shows me Vatoa reef, a long low silhouette of mangrove-enclosed shallows, a favored fishing place. Invisible beyond Vatoa is the great sea reef, and then the sweep of unbroken sea to the equator.

I have the sense that if something good is to happen between Malé and me it will happen here, away from the confusions of the village and the constant interruptions of a big family. Here we might construct a life that can harmonize our great cultural differences. I want to live on this island. It is so seductive.

Malé's father has told us that if we marry he will give us Vedrala. As the head of the *mataquali*, the extended family of his brothers and sisters and their children, he has the power to do that. He cannot literally give us the island, for it belongs to the *mataquali* and cannot be sold; but according to Fijian custom, whatever we clear and use belongs to us. The rest of the family has neither the resources nor the desire to leave the security of the *koro* for an uninhabited island.

The rain has stopped. A low red sun has broken through the clouds over Yaqaga, casting pink-gold light over the land. Malé looks at me, his dark face glowing, "We go?" We go.

CHRISTMAS

Christmas on Galoa was lovely. There were no winking colored lights, no presents heaped under decorated trees; but there was a sanctifying sense of joyfulness and peace. Everyone came home to the *koro* for the holidays, which lasted more than a month. Christmas is a time when families are reunited; sisters, brothers, children, return to their birthplace to celebrate the birth of the Christ child and the birth of the new year. Malé's brother Sione came from Taveuni bringing bags of flour, sugar, and rice; *yaqona*; and tobacco. Momo built a *vatu nu loa*, a big temporary open house with a thatched roof supported on poles, in which the family would gather. There meals were served to the entire family and any others who wished to join, and *yaqona* was drunk nightly.

Early on Christmas Eve, Momo called Malé and me into the

eating house. The whole family gathered. Malé and I sat on one side of the *tanoa*; Momo, Nei, Sione, Viliame, and the girls on the other. A freshly pounded heap of *yaqona* waited in a clean enamel bowl. The cloth for infusing the *yaqona* lay spread out across the *tanoa*; a bucket of clean water stood nearby. The polished, much-used *bilo*—one for mixing and one for serving— lay next to the basin containing the *qona*. All was ready for something important.

Momo began by praying. I could not understand the words, but the sense of holiness was apparent. He prayed openly, holding his large worn hands up to the heavens, speaking upward to the God that is his constant companion. When the prayer was finished, he spoke quietly to Nei; and she brought out a *tabua*, a huge old yellowed sperm whale's tooth, hung on an intricately woven *magimagi* cord made from coconut husk. Momo kneeled in front of us, holding the *tabua* out in front of him, stretching the cord and displaying the tooth. He spoke, and Malé translated. "This *tabua* is for you and Malé. It is a very old and very large *tabua*. We have kept it in our family for a long time. *Bula re* (good life), *maqa leqa* (no problems); love one another, protect one another, follow the God's way. It is our present of love from the family. We hope that you and Malé will be married. Keep this *tabua* as a token of our love for you and for your promise."

Momo held out the *tabua*, kissed it, and handed it to me. I looked at the heavy tooth in my hands, caressed its polished surface, thought of its journey, from a once-living whale to my lap here in the lamp-lit Fijian night. Momo clapped three, four, five times, the *cobo*, and said, "*A vūra*," drawing out the vowels into a spine-warming chant.

"Yeaaaa," the family responded, and clapped.

Malé said, "*Vina'a, vina'a va'a levu*," and I added, "*Vina'a* (thank you)." They were still the only words I was sure of.

The *tanoa* was brought into the room. We rearranged our seating: Momo, with Nei on his right, and Malé, with me on his right, faced the *tanoa*. Qare and Penioni, cousins, came in from outside, where they had been waiting, and seated themselves on either side of the *tanoa*. Sione sat in the center behind the *tanoa*, and Vili kneeled in front of Qare, a little to the side. The girls arranged themselves behind Sione.

This was to be *yaqona va turaga*, the *yaqona* for the chief, a formal ceremony. Momo occupied the position of the chief, and the *magimagi* cord attached to the *tanoa*, with a white cowry shell tied to its end, was laid out pointing to Momo. Nei, on his right, acted as *mata ni vanua*, the spokeswoman for the chief, and called the instructions to Sione for mixing and serving the *yaqona*.

Sione, Qare, and Penioni clapped three times, strong cupped handclaps in unison that resonated in the small room. Nei, as the *mata ni vanua*, said, "*Moli* (it's good)," and clapped three times. Sione then gathered the dried root in the cloth, twisting it into a pouch, and placed it in the *tanoa*. Qare added one *bilo* of water from the bucket.

When the *yaqona* had been piled in a small heap in the cloth and moistened, Sione said, "*Buli va ca tubea qona va turaga* (the yaqona is heaped for the chief)."

The *mata ni vanua* said, "*Tuwati loba* (mix the qona)," and Sione kneaded the yaqona with the water, with Nei calling "*Wai*" for each *bilo* of water that was poured into the *tanoa*. The *yaqona* was kneaded until the mix of water and infused root was the correct strength, Sione filling *bilo* after *bilo* of *yaqona*, pouring it from high into the bowl so that the *mata ni vanua* could judge, by the color, the proper mixture.

When the *yaqona* was mixed to the satisfaction of the *mata ni vanua*, Nei called, "*Talo qona va turaga* (pour the *yaqona* for the chief)." She then clapped three times. Sione ritually wiped the bowl with the bunched cloth containing the spent *qona*, holding his arms out in front of him over the *tanoa*, first a sweep to the right, then to the left; then he put the cloth into the basin and clapped three times. Vili squatted, bent-kneed, to the side in front of Qare, and Qare ladled one *bilo* of *yaqona* from the *tanoa* into another *bilo*. He handed it to Vili, who took it with two hands. Vili then straightened his body, brought both hands containing the *bilo* up to his chest, and extended his arms out stiffly to the front holding the *bilo* at shoulder level. He walked the few paces to where Momo sat, kneeled, and, with outstretched arms, offered the *yaqona* to Momo and clapped three times. Momo clapped once, took the *bilo*, and drank it down without stopping until it was empty.

He said, "*Moli*, *moli* (thank you, it's good)," and clapped three times. The *yaqona*, in the same manner, was then offered to Nei, Malé, and me.

When the four of us had been served, Nei said, "*Dou cobo, dou somi a veiqaravi* (you clap, and you who prepared the *yaqona* drink)." Vili, Sione, Qare, and Penioni together clapped three times, and a *bilo* was served to each. The *magimagi* with the white cowry was rolled up and placed under the *tanoa*. The ceremony was completed.

After the ceremony the family settled into an informal social-drinking session that would last all night. Other cousins, uncles, and aunts drifted in, and the conversation was noisy and happy. During the din I quietly excused myself and slipped outside. I sat on the clean wet sand near the outgoing tide, listening to the babble of voices. For a few cherished moments I had felt like a member of the family. Now I felt like an outsider again. It was bewildering.

Later that night I lured Malé away from the *yaqona* and gave him my promise in return, a handmade knife with a carved whale-tooth handle in the form of an eagle. It was the only thing like a *tabua* I had. There was no turning back. I wanted to see this through to the comfort of familiarity. I thought of Vedrala; it floated, unseen, the island of my dream.

Just before midnight we were all sitting in the *vatu nu loa*. The *tanoa* had been filled repeatedly; there was *qona* enough to last until dawn. Momo was leading the family in hymn singing and Bible reading. He asked each in turn for a favored passage and would then follow it with an appropriate hymn. When it came my turn, Malé asked me for some words. All I could remember were "The Lord is my shepherd, I shall not want." The family sang the psalm for me in Fijian. While they were singing, a wraithlike group of small children garbed in homemade white robes appeared carrying kerosene lanterns. They stood outside the *vatu nu loa* under the stars and caroled "Joy to the World" in a foreign tongue. A deep pounding from the *lali*, the Fijian hollow-log drum, echoing ancient rhythms; from the pastor's *bure* came the clanking of iron on iron. The church bells, *lali* and iron, were calling the *koro* to midnight service. Christmas had come to the village, a smooth blend of old and new ritual to assure that the world was at peace.

KORO DAY

Fiji is two tiny green globs on a transparent beach-ball globe I had brought from Hawaii to show the family something of world geography. It swings from one of the roof supports of the eating house and is a favored punching ball for Malé as he comes and goes. The two green globs on the globe represent the two largest islands: Viti Levu, big Fiji; and Vanua Levu, big land. There are some indeterminate specks that serve to stand for the other three hundred and fifty or so islands that compose the nation. Galoa, this island, is definitely not on a map of the world. She is one of many small islands of the north coast of Vanua Levu. One hundred and ninety acres, she shares a twenty-line paragraph in a geography of Fiji with her sister island, Tavea. At most a hundred and fifty souls come and go in the village. The geography says it has never been any different.

Everything outside of the village is called the bush. It is all divided into family planting grounds, or *teitei*. Each *teitei* is handed down from father to sons. Momo has his *teitei*, Vili his, and Malé his, although most of our food comes from Momo's garden; the boys are more interested in fishing than planting. Cassava, or manioc, is the staple, the bread of life, for it grows in indifferent soil, with little attention other than initial planting and weeding.

I struggle against the heat, flies, and humidity and follow Momo into the bush to his *teitei*, which is named Namoi. We follow the worn path that leads off the sand beach up into the bush. Momo is barefoot, as always, and walks with a splayfooted sure gait, his back bent forward slightly from years of working the soil. He carries his *sele levu*, machete or big knife, over his shoulder in a careless way that terrifies me, for the *sele* is razor sharp, honed to perfection by constant filing. Everybody has and carries one of these knives, including women and young boys and girls, and all are adept at using them. We are going to get the cassava for the day's food. After a few hundred yards the path dips down to a small white-sand beach, a soft curve of gentle bay, and leafing out over the beach is a giant *baka* tree. This is Wasa, the family burial ground, a tiny amphitheater of tropical garden with clumps of giant bamboo; spiky *voivoi* (pandanus) trees, the leaves of which are used in making mats; *avia* (mountain apple) trees; *tivi* (Fijian chestnut); and big breadfruit trees with dark green leaves and large pale green globular fruits. The graves themselves are clustered closely under the *baka*, raised rectangles of rocks filled with sand. One is a two-foot-high slab of cement with the name Jone Varawa, Momo's father, incised in crude letters. Near it is a small grave, the resting place of little Jone Varawa, Momo's first son, who died in infancy. Wasa is a cool sanctuary, imbued with the peace of the dead.

We continue a short way down the beach and climb to the saddle that separates the two hills that comprise Galoa. Garden plots have been hacked out of the head-high grass that coats the saddle. It is this grass that must be cut down with the *sele* each year in order to plant cassava or yams. The grass, curiously enough, came to Fiji as packing for ammunition for American rifles during the Second World War. Someone had the bright idea that the feather-soft seeds would protect the bullets and, without knowing, increased the labor of Fijian planters one hundredfold. Now the grass is well established in the islands and is sometimes used as an inferior roof thatching.

Momo's *teitei* is lush and healthy. It attests to his constant effort. The cassava plants are thick, the roots enormous. He loosens the soil around the cassava plant with a digging fork, kept in the garden, and then pulls up the entire plant, exposing a cluster of big heavy roots. Chop, chop with the *sele levu*, and the branches are thrown to the side to provide cuttings for the next planting. Fifteen or so roots are packed into a coconut-leaf basket, which he plaits on the spot. He then chops a carrying stick from the coconut midrib, slings the burden onto the stick over his shoulder, and we start home. The sun has swung almost halfway to the zenith, and it's really hot. At home Una has boiled the tea, and we sit in the stifling eating house, swatting at flies and drinking our tea. Later Suli or Una will peel the tough outer layer of skin from the cassava roots and wash and boil them in a big pot for the food for lunch and dinner. If there is fish we will eat fish and cassava; if not, then it's tea and cassava.

Malé has disappeared into the intricacies of *koro* social life. He is probably sitting, smoking Fijian tobacco, and telling stories with one of his many cousins.

Almost everyone in the *koro* is related. To one side of our family compound is the compound of Aunty Amele, Momo's father's brother's daughter. With her lives Qare, her nephew; his wife, Seine; and their two children. Seine is one of my best friends in the *koro* for she speaks English and has some insight into the ways of Europeans, being part European herself. She is quiet and careful, a graceful and kind person. Her husband, Qare, is inordinately handsome and one of Malé's singing buddies. Malé, Vili, and Qare have sung together so long that their voices blend perfectly. They and Uncle Jo, who plays a fancy lead guitar, were the quartet that wooed me the first week I was in the *koro*. On the other side of our compound is Ame's eating and sleeping house. Ame's mother and Nei's mother were sisters. He is the mayor of the village and an incessant *yaqona* drinker. Past Ame's compound are Uncle Luke and Aunty Una, related through Nei's side, then Aunty Una's brothers—Veresa, Saimone, and Jo—and their families. And so it goes, families melting into one another, throughout the village.

I prowl around half-heartedly looking for Malé in the heat. I long for a cool quiet place to lie down, but the eating house is full of chattering family, and there is nowhere to go.

It is often hard for me not to be angry and judgmental. The heat sometimes seems almost impossible to take; my mind and body rebel against the dirt, the constant discomfort, the flies. The scolding voice of Malé's mother wakes me in the morning and often puts me to sleep in the night. I think dark thoughts about Nei and then remember that her life is hard, much harder than mine; and what seems to me to be a lack of grace is just her way, born of the harsh and demanding life of rearing a family of three boys and six girls in a village where everything must be done with a maximum of labor, from making the morning tea to weaving the family mats or building the houses.

A day in the life of the women in the *koro* goes something like this: Get up before dawn, or at dawn. If there is water in the bucket or five-gallon plastic tank, fill the teapot and start the fire for morning tea. If there is no water, walk to the tank, or if the tank is dry, to Nabau, the distant well one-quarter mile down the beach, and carry back five to eight gallons of water hung on a pole over your shoulder. Wash the tin plates, cups, and spoons and scour the pots from the night before. Set the table, or if there is no table, spread a clean cloth on the mat and arrange the dishes. Boil the rice or set out the biscuits if there is rice or biscuits. Have tea with the family, chat, and tell stories about relatives and life in the village. Clean up the dishes and wash them. Clean the *bure*, sweep the mats, pick up all of the clothes lying around. Go back for more water and hand wash all the clothes; hang them to dry on the clothesline. Sweep up the compound around the eating house and sleeping house with a short broom made from the midribs of the coconut leaf. Peel the cassava and wash it and set it to boil. Go into the bush and scrounge up firewood. Come back, carrying a heavy load over your shoulder. Drop the firewood with a tired clunk next to the fire. Clean the fish, if the men have brought fish, and fry it, if there is oil, or boil it if there is none. Set the table again for lunch. Serve the food and eat what is left after the men have finished. Rest briefly and perhaps then go back into the bush and gather *voivoi* leaves for future mat making. Go back to the tank or well for more water. Come back and boil the tea. If there is time and the tide is right, perhaps go out on the reef in search of small fish or shellfish. Come back and cook whatever is available for dinner. Maybe grate coconuts to make coconut cream, or *lolo*. Wash the dishes again. Set the table for dinner. Fill the kerosene lamps and light them. Eat and tell stories. Clean up. Carry a bucket of water to the bathing house for your husband or brother. When he is finished bathing, go get one for yourself and bathe. Rest, or sew, or weave mats, or maybe grate twenty or thirty coconuts for coconut oil, or stoke up the fire and bake buns or bread for the morning tea. Go to sleep at nine or ten o'clock.

The day varies, of course, according to the season, the tide, the necessities of *koro* life—which might require making large numbers of mats or feeding visitors for a ceremony—and the size of the family and the number of women sharing the labor. Some husbands and brothers help: carrying firewood and fetching water or coconuts from the bush. Others help not at all. This sketch does not include taking care of small children or infants, although usually a woman is allowed to rest when she is caring for a newborn baby. Usually, but not always. And this schedule holds for pregnant women, many of whom work like this until the time of delivery. The women are extraordinarily strong and good tempered, laughing and making light of their constant labor. But it is constant, relieved only by Sunday's minimal rest.

The men work hard too. Their work is intense, requiring great strength and endurance. They weed, dig, and plant in the gardens; go to the *teitei* for the day's cassava, pick the *ulu* from the breadfruit trees, build houses, and fish. Often they will fish all night and return with too few fish to pay for the gasoline (called benzine) to get to the fishing ground and to Munci's store, about a half hour up the Lekutu River to the mainland. Sometimes the men catch a lot of fish, and it is too late to sell the fish, or Munci won't buy them. Surplus fish that are not sold are shared or smoked over an open fire for future use. Smoked fish will keep three to four days. No fishing is allowed on Sundays, so Friday or Saturday is the time to smoke fish for the Sunday lunch.

Here in the village there is practically no money. No one, except the head of the household, has any money at all; and a big family might spend only five or ten dollars a week. The only things bought at the store are flour, oil, rice, sugar, tea, benzine, kerosene, *yaqona*, and cigarettes or Fijian leaf tobacco. It seems somehow there is often money for *yaqona* and tobacco, if nothing else. The money, what little there is, comes from fishing and goes to pay for the benzine to do the fishing, the white gas for the gas lamp, and, if you are lucky, a two-kilo bag of sugar and a package of tea.

In the late afternoon I walk to the village store with Ima to buy some cooking oil. Since I am "rich," I can buy a full bottle, about $1.60 for 750 milliliters; most of the time it is bought by the cupful, and the kerosene is sold in empty quart beer bottles for 40 cents. The store, a tiny tin shack with a wooden counter and three shelves, is stocked with three bottles of oil, a half bag of rice, a half drum of kerosene, five packets of biscuits, five packs of tea, and six small tins of fish. It is run by cousins Tovi and Penioni as a community service, for they buy their supplies at Munci's store, and I doubt if they make enough money to pay for the benzine used in getting their stock. But it is a cheerful place, and Tovi or Penioni will always stop what he is doing, open the store, and carefully note your purchase in a worn copy book. They are both very precise about the change.

Apparently the Fijians inherited from their English governors a great fondness for tea with sugar, and they drink it constantly, sweating in the humidity, making fire all day long to boil the ever-present kettle. If I walk through the village on an errand, there is always the calling out of voices from the different *bure*, "Joana, *somi ti, somi ti* (a call to come and drink tea)," which etiquette requires; and an etiquette just as strict demands a ritual answer, "*Sa re, vina'a va'a levu* (thanks, thank you very much)." Whatever my mood, whatever my mission, I must conjure a smile and a thank you. Rarely do I stop, rarely does anyone stop, but not to offer would be a great breach of manners, just as at mealtime one is invited to come and eat by everyone whose house one passes. And it is impossible to walk through the village without passing all of the houses and without being seen by all the people looking out through their doorways at whatever moves in their field of vision.

So there is tea, tea, tea, until my mouth longs for something cold and sour. I can't understand why, with coconut trees all around and men capable of climbing them, they do not drink the cool fresh water of the coconut, but choose instead the hot sugared tea that must be bought with very hard earned money and boiled at the cost of collecting and carrying great bundles of firewood and squatting by the hot smoky fires in the middle of an unbearably hot day.

Thank goodness night comes and the flies go to sleep. Uncle Saimone's son Kavaia has brought over some fish, and after a noisy lively family dinner of the cassava and boiled fish, Malé suggests that we go to the other side of the *koro* and drink *yaqona* with his Uncle Paula and Aunty Monomono. Uncle Paula is Tevita's brother, an uncle to Desele and distantly related to Malé. He is educated, and he and his wife, who is a schoolteacher, speak fluent English. The invitation sounds good to me, for I will be able to participate in some conversation instead of sitting a bewildered stranger in the swirl of Fijian stories.

A soft dark starlit night, the moon will rise late. Malé disappears into the blackness ahead of me, impatient to be on his way. I call him back, "Please come walk with me; I can't see where I am going." He returns with reluctance because he is in a hurry to get to the *yaqona* and because he is embarrassed to be seen walking with me.

I choke back my irritation, remembering that Malé is new to the exotic ways of Europeans, ways that include husbands and wives, or lovers, walking close together, and stumble along in the darkness trying to keep up.

Uncle Paula's big wooden house is crowded. Almost all of the men in the *koro* are here, lured by the dull pounding of the *yaqona* in the *ta bili*. There is no way a grog party, as they call a *yaqona* session, can be kept secret in the village, for the sound of the *yaqona* being pounded reverberates through the earth, a deep and unmistakable summons. Monomono is sitting near the door, and foregoing my usual insistence that I sit next to Malé, I take a place on the mat next to her.

She and I are the only women present. "Hello, Joana, how are you?" The familiar gracious words fall like cool water into a willing basin.

We chat, about this and that, how Malé and I are getting along, my problems in making my needs understood and accepted. "You must teach him," she says. "You must tell him every day so that he can understand. He is very young; he has no experience with European ways; you must be patient. It will take a long time."

Soon we exhaust the subject of the difficulties of an intercultural love affair. Besides, the *yaqona* is getting to me, and what seemed impossible this afternoon recedes as a minor inconvenience. Our conversation turns to Fijian customs, an intricate maze of propriety that varies greatly from place to place, village to village, much as the language varies.

The foundation of this maze is the *mataquali*, the extended family usually headed by the eldest male. This is the landholding unit of shared responsibility and support. If Malé's father needs something, he can ask it from any member of his *mataquali*, and his wish will usually be honored. With this system of mutual aid goes the *kerekere* system, an accepted form of begging for anything one might need, from a cup of sugar to the use of a boat or the gift of a cherished new jacket. One can say no, but more often than not, one says yes.

Uncle Paula has been listening to our conversation and adds that historically Galoa was composed of seven *mataquali* with hereditary roles. The first was the *turaga*, the chiefly *mataquali*, from which the chiefs are selected. The *sau turaga*, the second in importance, had, however, the right to name the chief. After that came the *mata ni vanua*—the face, or front, of the land—the chief's spokesman and herald. Momo is the *mata ni vanua* of this *koro*, and on his death Vili or Malé will probably assume the role. Next was the *bati balavu*, the warriors, whose ancient duty was to protect the chief and fight according to his wishes; then the *gone dau*, or fishermen; the *bati lekaleka*, or gardeners; and finally the *mātaisau*, the carpenters and builders. In the traditional village structure these roles were strictly followed and resulted in specialization and a high degree of skill. As the chiefly system breaks down, so do the other traditional roles, and, sadly, so does the skill. In the old days one was expected to marry within the *mataquali*.

A woman, upon marriage, is expected to go to live in her husband's village and become a member of his *mataquali*. There are exceptions; sometimes the wife's *koro* is more advantageous, but as a rule the custom is followed. If I marry Malé, I will become a member of Nalomolomo, his *mataquali*. Each *mataquali* has a hereditary *vū*, an ancestral fish or animal whose flesh was once *tabu*. The word *vū* means basis, root, the source, the place one came from, and is found in compounds such as *yavu*, the house foundation, which is also handed down from father to son. The *vū* of Malé's *mataquali* is the *dadakulaci*, the black-and-white-striped sea snake. It is poisonous. Uncle Paula's *vū*, the *vū* of the fishermen, is the shark.

In the old days the chief's word was law; he was considered to possess great *mana*, or supernatural power. I look across the room at the chief of Galoa, a small unprepossessing man who does not appear to possess much *mana*. The *mana* and power of a chief rest with his hereditary status and the great high chiefs of the historic centers of power; Bau, Rewa, Somosomo, Bua, Lau are still accorded awed respect in the nation and possess great *mana*. The chief of Galoa is a minor chief of a small *koro*.

I look at the *tanoa*. It is half full. This is the last round, the *kosa* (dregs) have already been pounded. Almost everyone is semicomatose. The noisy conversation has subsided. Everyone still sits, for it is the custom not to lie down or recline while drinking, but most are half asleep and have to be prodded awake to drink the next *bilo*.

Monomono smiles at me, "Goodnight, Joana, I'm going to sleep."

"*Moce*, Mono (goodnight)."

I wait patiently until the *tanoa* is empty. Qare tilts the empty bowl forward, "*Sa mada yaqona va turaga* (the *yaqona* of the chief is empty)," he says, slurring the words together. He wipes the bowl; stacks the *bilo*, cloth, and basin into the *tanoa*; and sets it to one side.

There is a short pause, and then everyone stirs, gets up, and leaves saying, "*Vina'a va'a levu yaqona, vina'a va levu* (thank you for the grog, thank you)." It is 2:30 A.M.

Malé and I walk through the dark quiet village. The moon still hasn't come up, but the stars are brilliant. When we reach home, everybody is asleep. He rifles around in the big pots and comes up with a few pieces of cassava and some boiled fish, eats quickly, puts the dishes outside on the *vata* (shelf), and gets into bed. I follow. In less than a minute he is asleep. A day in the *koro* has ended.

WHY?

Malé just returned from fishing with his sister Va. Diving to thirty feet with the new speargun I bought him, a marvelous and cherished tool, he had had a bad nosebleed and got scared and came home. Now he can't hear out of one ear, and there is no fish for dinner, so we will eat rice and breadfruit and drink tea. Tomorrow they will try again.

He is asleep now, refusing my ministrations in a stolid uncompromising way. Fijians are not much on medicine, preferring instead to let time and fate take their course.

In a spirit of foolish bravado and clear-cut jealousy, I went fishing with Malé and Vili for three days last week—to have something to do and to be with Malé, instead of watching him go off with his competent and hardy sisters and brothers, leaving me in a fit of hurt feelings. I paid dearly for my childishness, for now the back of my head is covered with boils, the product of heat and lice, and aggravated by sunburn, and I have had to chop off all of my dearly loved hair. Cutting my hair was an act of some commitment, but it was either go home to Hawaii with my hair and leave the dream smoldering as a dream, or commit myself to Fiji, hurricane heat, and lice, and cut my hair. As soon as I cut my hair, I felt better, as much for the coolness as for the release from the constant feeling of not being able to cope. Besides, everyone says I look much nicer, and I can use a fine-toothed comb to get rid of the lice.

Now it is clear that the sweet image of Malé and me fishing together will have to wait until the cooler season, if one will come, and I must make a further adjustment in my image of the tropical paradise.

When I first came here and Tevita forbade me to go to visit Malé, saying he was a very bad person and wanted me only for my money, I was suspicious, for didn't Tevita and his family also want me for my money? All of their stories about his badness whetted my curiosity. I have always had a fascination with the semioutlaw; and the things Malé had done, mainly getting drunk and fighting when he was staying in Suva, didn't seem that bad. I was also a little seduced by the possibility of helping him to change his ways, which he had told me he wanted to do.

So, much like the moth to the flame, I flew unhesitatingly to the danger. There was not much basis for the relationship except that Malé wanted to marry me, and I was intrigued by him. Malé is young, and although very skilled as a villager, has no comprehension of what I would consider a love affair or marriage. His idea of a wife—the Fijian idea—is basically another sister you can order around, yet one you can sleep with. We could hardly speak together. Yet the silence, too, was refreshing, for if we couldn't much speak, we couldn't much get into trouble with one another. So we collided like two mute planets in the night, in a quiet explosion of dark and light.

I have always been fascinated and attracted by darkness, by the soft dark night, by the shadows, by the depth of dark eyes. Dark skin seems more functional, and more attractive, to me than light skin, for it doesn't burn in the sun and get lumpy and red, but stays clean and beautiful. My early arguments with my mother, which continue to this day, have to do with my attraction to dark men. So here comes this handsome Pacific Islander who is said to want my money. And what does he have to offer? An island, the dream of beauty, and the possibility of love—and how much is that worth in an age when even a cheap car costs five thousand dollars?

Yesterday I received a letter from my mother. She is very upset. She is old, and I am going farther and farther away. She writes, "We can't understand what you are doing out there, why you have left the soft green paradise of Hawaii, your house and horse, and dog and car and boat, and thrown it all away to go live in a village." Well, Mother, I can't exactly understand it either, especially when there is nothing to eat but boiled breadfruit and when sometimes even the act of getting a drink of water is more than I can manage. I'm sure if you saw me here you would understand even less. Then why?

I once met a man—accomplished, educated, affluent—who told me that he had read in a book that if one does not follow one's dreams, one dies. No matter how long the body stays alive and well, the spirit withers, the soul departs, leaving only an empty shell, a husk lacking psychic vitality. I think it is something like that; the dream is the seed, and if you kill it, it will lie inside like a child never born, poisoning the body and the mind, a rotting clot of something that might have been graceful and beautiful. I'm not sure it matters whether the dream can ever be real, yet the way must always be open for it to materialize.

Three days before leaving Hawaii to come to Fiji for the first time, I spent a sleepless night in the anguish of a full moon. Lying alone in the familiar safe comfort of my bunk on the boat, I longed for a warm man at my side. I wanted to sleep in a grass house, to lie next to an islander under the South Pacific stars. That's what I wanted—not a new dress, or car, or a date to dance in a fancy hotel, but a grass house and the soft sound of the sea and the great dark night, and a great dark man.

The morning light filters into the *bure*, a pattern of stars woven into the coconut matting. The black-and-brown-and-white pattern of the *tapa* hanging over our bed stirs into life; the mosquito net flutters in the breeze. I wake and look at Malé sleeping beside me. Stretched out full and comfortably, his face relaxed, his long lashes covering his intense dark eyes, a bright red *sulu* glowing with huge orange hibiscus flowers thrown over his loins, he is the dream incarnate.

The morning birds are stirring, a soft calling of doves and honey-eaters. I lie next to Malé, warm with his warmth, and remember my longing.

FIRST HOUSE

Yesterday we went to Vedrala to work on the little *bure*, or sleeping house, Momo is building for Malé and me. It is a tiny house, no more than ten by twelve feet, with a sand floor which later will be covered with a thick bed of dry coconut leaves and a mat. The walls and roof are of coconut matting. Malé, Una, and small Jone went out fishing; Momo, Ima, and I stayed on the island to work on the house. The week before, Momo, Nei, and the girls had cleared the ground, chopping up young trees and pulling up grass and weeds to clear a place for the *bure* under the big *rara* tree on the beach.

Momo is an intense, strong man, an indomitable worker, skilled in Fijian carpentry. Saplings are cut down in the bush to make the framework, a branch is sharpened for a digging stick, and holes are dug in the soft, sandy ground to set the poles in a rectangular framework, measured only by Momo's practiced eye. He then sets the first four corner posts, cut so that the crotch of the tree receives the roof framing. He nails four long poles into the crotches and adds additional supporting poles. In the old days, before such exotica as nails, the poles would all be lashed into place with *magimagi*, rope made from the coconut husk. Soon there is the skeleton of a house, a rectangle with a peaked roof. When Momo runs out of poles, he disappears into the bush and soon returns with a bundle of poles over his shoulder. His only tools are: the *sele levu*, to cut the branches off the saplings; a hammer; and a small ax, to cut down the poles. Lacking the ax, he could build the whole house with the *sele levu*. When the framework is complete, he goes down the beach and climbs a coconut tree.

I look up and see him there, feet and knees clasped around the tree, high up in the crown, lopping off branches with a sure chop of the *sele* and dropping them. A great pile of fresh green branches collects under the tree, and the crown is a naked shivering thing of six or seven fronds. He climbs down the tree and, neatly piling up the branches, lifts eight of them onto his shoulder with a grunt and carries them back to the clearing. He is invisible under the long heavy swaying branches, a green-leaf apparition moving down the beach. When he has collected forty or so branches, he nicks the midrib with the *sele*, rips each apart with his hand, chops off some excess midrib, and sits down, his legs spread out in front of him, next to the framed house and begins to weave the *bola*, the mats that will become the roofing and siding of the house. While he weaves the *bola*, we talk, a mixture of basic Fijian, basic English, and gestures. We talk about work and food and God. And we talk about Malé.

It is amazing to me that we can communicate serious abstract questions of life and morality with this simple stew of words and motions.

"Malé is a good boy," Momo tells me. "You come live here on Vedrala. The *koro* no good, only go round, go round. Live here with Malé. Good, no live in the *koro*, no good. Too much *pāpāpā* (gossip). The God here, no *pāpāpā*, can think. *Wawa* (wait), Malé come good."

We talk about *leqa* (problems). There are *maqa leqa* (no problems) and *leqa levu* (big problems). If there is food—*ulu* (breadfruit), *bia* (cassava), *i'a* (fish)—there are *maqa leqa*. If there is no food, there are *leqa levu*. If there is land (*vanua*), there are *maqa leqa*; if no land, there are *leqa levu*. I think back to Sam Whippy in Honolulu and his "no problem." Now I understand where that came from.

Momo is lucky. His family has an island, this one, and he has his land in the village: his compound and the garden at Namoi, and another at Nalomolomo, where he plants bananas, more cassava, and yams. But, most important, he has his vast energy, his strong belief in God, and his desire to work.

Momo describes to me his father's death. His father called all of his sons and daughters to come to him when he knew he was dying. When they had assembled, he pointed to Momo, not the eldest, but the hardest worker, and gave him Vedrala. Momo tells me that he will give Vedrala to Malé when he dies, that Malé, not his older brother, is his heir. He tells me that the God sent me to Fiji to help the family, that he is getting old and it's harder and harder to labor. It's the God's way, you stay with Malé, marry, it's good. And we sit there companionably, under the great cool *rara* tree, weaving house mats of coconut fronds and discussing land and food and the necessity of work. I feel very familiar, as if this man is already my father-in-law, and it is natural and comfortable and right. At times like this it seems the most easy and appropriate thing to do: to marry Malé and make my life real under the floating shadows of the dream.

Momo tells me this small *bure* is only the beginning. Little by little. First a small house, then a little bigger one, and later he will build us a true *bure*, not a temporary *bola* cottage like this one, but a big substantial, traditional Fijian house with thatched roof and heavy hardwood poles, a house that will last.

This tiny house looks like a big basket, woven together, each piece weak, yet the whole structure strong and light. The *bola* mats are lashed onto the poles with rope made from stripping the bark off of long straight saplings—*vau* trees, or a vine called *tuva*. The process is called *cori bola*. He teaches me how to tie the mats onto the framework, and I am made to repeat the words over and over so that I can tell Malé on his return from fishing that I was *cori bola*!

When we leave Vedrala late in the afternoon, the *bure* is three-quarters complete: a fresh green doll's house set in a clearing under a great tree.

HURRICANE SEASON

Somebody named Kenny sent us his calling card last night. It came in the middle of the night in a flurry of driving rain and in a sky that flashed and glared and rumbled. He had given us warning that he might visit: three days of terribly hot humid weather; and Saturday, when Malé, Vili, and Qare went fishing, there were no fish hiding in the coral—it is a sign that heavy weather is coming when the fish head out to the open ocean—and then the current changed and became muddy. So before the radio announced the presence of a moderate cyclone to the southeast, the fishermen already knew of its approach. The day we were on Vedrala weaving the *bola* Malé's father told me that the heat meant bad weather is coming.

Last night when the sky opened up and fell on us, we all woke. Malé and I, and his sisters sleeping in the other bed, dragged our mattresses away from the walls and windows. For some reason the downpour caused us all to want to pee, and as I squatted in the waterfall in the doorway, I looked around in the dim light of the kerosene lantern to see Malé peeing out the window and Una peeing out the other door. It seemed ridiculous, adding our trifles to the deluge, but it made everything manageable and a bit funny. We all settled back into bed, and Malé put his arm around me, a thing he rarely does, and I curved into his warmth and listened to the deep-throated roaring of the heavens, looked out at the flashing glaring light opening up the sky, and felt happy, warm, and secure.

So far it's not really as bad as the radio makes it sound, but if the hurricane comes directly this way, we will all go into the new concrete-block church, a true shelter from the storm, and sing hymns and pray and draw comfort from God and one another. But that is not yet now, and now there is the stirring lurking presence of this uninvited guest, Kenny, some two hundred miles to the southeast, and the reassuring placement of our belongings inside the *bure*.

On Friday Bill, who is building a new boat for us over on the mainland, came and asked if he could use his old boat, which we had borrowed while awaiting completion of ours, and our brand-new 40-horsepower outboard. He had promised to take some friends fishing on Naniqaniqa, a small island a few miles beyond Vatoa reef. Malé was not at all happy about it, but I could see no way to refuse gracefully; after all, it was Bill's boat, so we lent it and our new outboard to him for the weekend. True to Fijian style, Sunday afternoon has stretched out to Tuesday morning, and still no sign of Bill, boat, or engine. For the last two days Malé has been staring pensively to sea listening for the sound of the return of his cherished engine. This morning, in response to his anxiety, I told him to hire a boat from Galoa and go find our engine.

Malé and Vili are gone now, and I am hoping that this shifting calm holds long enough for them to return safely, and I am feeling bad about not going along. Danger experienced is easier than danger imagined, and I am becoming so enamored of Malé now that I cannot imagine losing him.

The village is preparing for Kenny. The men have returned from the bush, where they cut coconut branches to lay over the roofs and secure the *bola*; and a sound of hammering is heard in the heavy calm that precedes the storm. Everything that can blow around has been brought inside; the pots and dishes that usually dry on the *vata* in the sun are in the kitchen; a small supply of firewood is stashed in a dry place. I have made some ineffectual attempts to put away my things, hiding the ukuleles and my typewriter under the bed and putting my writing and clothes in a suitcase—all of which will be meaningless, I know, if we are visited by Kenny in all of his indifferent tropical splendor.

The *bure* look strangely festive in the gusting wind and lowering sky. The fresh green branches of the coconut contrast with the dried brown of the *bola*, making the rooftops look as if they were decorated for a celebration instead of readied for the threat of ninety-mile-per-hour winds. I walk through the *koro*, watching the preparations for the coming storm. The freshness of the wind, the gray heaving sea, are comforting. Everyone seems cheerful and calm. This is not the high-pitched voice of the radio announcing some faraway possibility, but the presence itself of the wind, the heavy sky, the black clouds gathering.

Malé's father just called out, "Joana, the boat is coming. There. Malé, coming." So my fear for Malé must be filed away in the dead-letter file of useless fears, and we will smoke a cigarette, listen to the story of finding the boat, and I can look into his unclouded happy eyes; the engine and Malé, safe.

When I thought of coming to Fiji during the hurricane season, I thought only of the symbolic possibility of hurricane. Primed by romantic storm stories, I thought it might be fun to watch the coconuts bend to the earth, to feel the power of nature. The hurricane season is not, however, the refreshment of storm, but the heavy hot waiting for nature to declare herself. It is waking in the morning to the anguish of another day spent in the heat and infuriating buzzing of flies around my face, the feeling that I cannot stand another minute of this gathering threat. Day after day piles up sticky and dense, so that this day, when the wind blows cool and the gray sea billows, I feel strangely comfortable and alive.

I hear Malé calling from outside, "Joana, you want to go fishing with me on the reef?"—a curious occupation in the gusting wind and sheeting rain. I say yes, for today I have energy and interest because of relief from the heat. We drink a quick cup of tea, and he ambles around finding his spear. I put on a windbreaker as scant and unneeded protection from the rain and follow Malé, who is barefooted and wearing only shorts and a T-shirt, out onto the reef in front of the eating house.

Eager to be after the fish, Malé strides out in front of me. I pick my way through a zone of refuse that would delight a future archaeologist, for all of the village garbage is thrown on the beach to be carried away by the tide. Broken coconut shells, rotting cassava, fish bones, and shells attest to the basic diet. An occasional rusty tin signifies store-bought luxuries. Ah! fragments of a rusting kerosene lantern, a soggy tea package, a mangled coconut-leaf basket, a disintegrating turtle shell offer further insights. No broken bottles and a lot of sea-brought rubbish contribute to the puzzle. After about twenty yards I am out of the village tidal zone and onto a stretch of muddy sand shallows: home of knobby fleshy sea creatures, worn dead corals, and a few anxious small fish. Malé is way ahead of me, almost invisible now, his dark brown body blending with the low brown rocky silhouette of the far reef. He will hunt the fish at the intersection of dead and living reef, where the muddy brown mingles with the pale greens and yellows of the living coral.

I quicken my pace, but carefully, for I am still uncertain on the reef, unlike Malé, who can run barefooted through coral chasing fish, and though never seeming to look where he is going, always knows. I cannot believe that he can actually see the fish in this dark water, for there is no sun to reveal the subtle glint of color that betrays his prey, but I see him crouching low, stalking, the spear ready. Malé hunts with a spear he has made from a 12-foot length of fire-hardened cane and four file-sharpened pieces of pencil-thin iron he has bound onto the cane with wire to form the prongs. The spear is light, and in his practiced hands, accurate but fragile. Once, in a playful mood, I jabbed at a rock with his newly made spear and promptly broke it—a lesson in spear handling I won't forget.

Seeing him fishing like this, in the gusting wind, outlined against the gray sky, intent and certain, he pops into focus as my dream savage, and all of my anger over his shortcomings melts into admiration.

Slowly we make our way around the reef that encircles the island. We round the point of Galoa nearest the mainland, no more than two miles away; and the wind freshens. It is blowing from the southeast, where Kenny, somewhere, rages unseen.

Malé heads farther out on the reef, for there, in the deeper water that separates Galoa from Vedrala, huddled unromantically in the monochromatic light across from us, he may find the bigger fish. In the distance we see another dark bent figure stalking, and Malé heads in that direction to find Lala, a distant cousin, who has just caught a *saqa*, familiar to me as the prized *ulua* (trevally) of my Hawaiian days. This is a big fish, tasty, with few small bones, and one can feed the family. Malé and Lala hunt for *saqa* while I wander the shoreline among the mangroves seeking shelter from the increasing rain.

I would like to be out there fishing with Malé and Lala, but *coco'a* (fishing with a hand-thrown spear) is a skill learned only with years of practice. First you have to be able to see the fish, not easy even under the best of circumstances, for the fish are well camouflaged; then you must stalk them without their seeing you; then you must hurl a spear anywhere from five to twenty yards and hit a small moving target, while making an allowance for the distortion of the water.

After a half-hour's unsuccessful hunt, Malé and Lala head my way. Malé is carrying his string of smaller fish gleaned from the two-hour hunt, and Lala is carrying his *saqa* on the end of his spear. We cut up into the bush, climbing an almost imperceptible path through a tangle of vines and high grass. A gang of young boys is happily and noisily throwing sticks at a passionfruit vine to drop the fruit; and Malé, always ready to join the fun, runs ahead to join the game.

I find a suitable rock, lay the string of fish at my feet in the glistening emerald green leaves, and watch the sport from a safe distance. The passionfruit vine loops high up in the branches of a mango, the green fruit barely visible. All of the boys are accurate with the throwing stick, and plop, plop, plop, down come the fruits.

"See, Joana, see. Here," Malé comes my way, offering an armload of hard fruit. We crunch into the gathered fruit, picking our way through the bush; the hard sour taste is a welcome change from the constant diet of soft fish and rice and cassava.

When we reach the compound, the family admires Malé's catch, and his sister Vaseva cleans and fries the fish for a late lunch. I change into dry clothes; the rain has moderated into a soft drizzle, and I sit down with Malé, eating fresh fried fish while he tells, in intricate detail, the story of his hunt.

Malé's shorts are still wet, but he is hungry and doesn't care. I look at his body, salt stained and strong, and think of his chasing the fish in the shallows, leaving little puffs of spray behind him.

"Here, Joana, eat, eat," he says, piling the best pieces of fish, which he has just deboned, onto my plate and squeezing fresh lemon juice onto the pile. " *'Ana va levu* (eat a lot); it's cool now, you can eat." My appetite has fallen drastically because of the heat, and I am fast losing weight. Malé pokes me playfully in my shrunken stomach—a hard poke—and continues his story.

After lunch Malé tells me he wants to take the boat to Nakadrudru, some four or five miles away, to Munci's store to buy sugar, for the family is out of sugar. It seems crazy to go out on the sea during the threat of a hurricane to buy, of all things, sugar; and I say no, no way am I going to risk my life for two kilos of sugar. After a while I reconsider and think, why not, for what is life if I am always saving it for later, and a trip alone with Malé will provide a diversion from the constant chattering and my idleness in the village.

The sea is gray and covered with whitecaps, but not as dangerous as seas I have already traveled, and the sky and clouds have held steady since morning. The wind is blowing about twenty knots, not much, and the waves, though breaking, are not very big. Thinking about it, it seems that we will probably return alive, and since there is no food but breadfruit, and it is just possible that Kenny may come this way, it seems smart to go to the store for some flour and rice and tinned meat. We get into the boat under the watchful eyes of the family, put on wet torn yellow slickers, and set out on this hurricane adventure.

The seas are not that big, but Bill's plywood runabout is old and frail, and as I watch Malé work the waves from the stern, unconcerned about the wind and spray, I realize that my life indeed does depend on his skill and ability. At times like this I forgive him, for I see that beneath his demands and remoteness is a skillful man.

Malé is very good at this game, smiling and joking and playful as he shepherds the fragile boat over the breaking waves. After about a half hour we reach the mouth of the Lekutu River, swollen and muddy, dangerous with floating logs washed down from the mountains. It is getting late, and if we are to reach the store, buy what we need, and return before dark, we must hurry. I don't want to come back in the dark, for our safety depends on seeing each wave. When we reach Nakadrudru, we hurriedly charge some groceries—tins of meat, powdered milk, coffee, cocoa, sugar, flour, rice—oil, batteries for the flashlight, kerosene and benzine for the lamps, and go back to the boat.

The tide is starting to go out again, and when we reach the mouth of the river, the shallow sea has roughened, the sky is a dark gray, Galoa a black hump in the dim light. The wind is on us now, and the going more tricky. I cannot stand in the boat and instead sit under the small cabin roof, clutching its sides, watching Malé pick his way through the breaking waves, alternately gunning and idling the engine. He is in his glory, intent, happy, and boasting of how good he is—and he is good, and I am proud and happy to be with him. But my calm and happiness are definitely because of our closeness to Galoa. If I had to spend several hours doing this, I would be terrified.

A crowd has gathered in the last light of the day, lured to the beach by the sound of our engine; and the whole family, including cousins, is there to wade into the water, steady the pitching boat, and unload our precious cargo. A few minutes later we are all sitting in the eating house in dry clothes, telling the story to an admiring audience. There is food on the shelf, enough for three or four days, and we are drinking hot cocoa and eating crackers with margarine and jam. This meager hurricane feast tastes delicious. Could I ever have imagined how good were cocoa and crackers, eaten while sitting on the mat under the comforting eye of a lamp that has kerosene in it, while Kenny flirts with our future somewhere outside?

⯭ ⯭ ⯭

Hurricane season. Gigantic forces are restless. All night the lightning flashes in a crystalline sky studded with stars. I have never seen lightning so intense in a clear sky, and it confounds me, light blocking and flashing in the inky blackness, blotting out the glory of the stars, and then returning it, silent and remote.

Each day dawns hot and uneasy, the clear sky slowly turning into a mass of dark clouds over the mainland of Vanua Levu, with the ever-present throaty rumble of thunder, the intensifying heat. When the sky has turned dark and enough water gathered into the clouds from the steaming sea, then the rain comes in an opaque curtain that stings the eyes and blots out islands and horizon. Sometimes when we are in the boat, the rain will come like that, a gray wall advancing, until we cannot see where we are, and I marvel at Malé's ability to find his way home.

"Where is Galoa?" I ask, "where is Galoa?"

And he laughs and says, "Why you can't see Galoa?" He thinks it's a great joke, appropriate for the season.

It is only during and immediately after the rain that I find respite from indolence and the heat. Most of the time I sleep, hot sticky sleep that mats my hair and leaves me feeling more tired on waking. The sleep is so dense and uncomfortable that I must sleep some more to avoid the discomfort. And so I fret because I am sleeping my life away, eating less and less, and getting thinner and weaker, and all of my grand plans for teaching Malé and the family different ways swirl and float, evading me under the mosquito net. It is only when the wind blows fresh and I feel again the cool energy of comfort that I wake in my mind and find this life possible.

LABASA

Labasa. Crowded, noisy, exotic. One long street is strung with shops offering all of the wares that can be bought on Vanua Levu. The shop windows display a bewildering array of freshly baked bread, bundles of *yaqona*, sweets, cookies, cheap plastic toys, Seiko watches, Chinese guitars, plastic basins, kerosene lamps, galvanized buckets, and colored lithographs of Indian gods. Entering the small hot shops, one finds sacks of flour, sugar, rice; shelves of tinned fish and corned beef, stewed tomatoes, and syrupy peaches; stacks of cigarettes—always stacks of cigarettes. There are building and hardware stores, fishing supplies and outboard motors, cheap plates of curried mutton and fish, and expensive pickups and refrigerators. The big open market is bursting with vegetables. Patient farmers, mostly Indians, sit for hours near carefully arranged and counted piles of lemons and oranges, eggplant, cabbage, and beans. They look hopefully into my face as I pick my way through looking for the brightest, the biggest, the freshest. Inside the great darkened shed are counters heaped with the unwrapped *yaqona* and the dusty sweet smell of future calm. Leaning against the counters are big sacks of rice and *dahl*, potatoes and onions. Some of the rice is grown locally; the rest and the potatoes and onions are imported from Australia and New Zealand, as is all of the packaged food in the stores. Tons of powdered milk, cooking oil, and white flour are unloaded in Suva, where they are stuffed into boxes, sacks, and bottles and sent out over the islands, deceptively granting the illusion of Fiji's self-sufficiency.

Labasa is an Indian town. The shops, houses, buses, taxis, and most of the cars are owned by Indians. There is a curious split of races in Fiji. East Indians were imported by the colonial British government to work as indentured laborers on the sugarcane farms. Fijians were no good; they didn't like labor uninterrupted by play, and they languished away from the familiar warmth of the *koro* and the psychic support of the family. Also, there was a definite danger that the native Fijians would starve if the men were forced to leave their fishing and gardening to produce sugar cane, so over the years about sixty-two thousand Indians were brought into Fiji to work the cane. As they completed their indentures, they leased small farms from the Fijian *mataquali*, and slowly the control of the rich, commercially productive flatlands drifted out of native hands into the hands of the Indians. Indians now control the sugar and rice lands. Seeing Fiji as a land of possibility, wealthy Indians have recently immigrated to establish large stores and manufacturing, import, and transportation businesses, so that the economy and a great deal of political power are in the hands of wealthy Indians. The Indians who control the economy aren't, however, the ones sitting behind the sacks in the market.

Most of the Fijians stayed in the *koro*, where they enjoy themselves, having an intense regard for one another and somewhat of an indifference to things: things are great, but they are not as valuable as a family. The Fijians who drift into the cities are often jobless, impoverished, and without the support of the land and the tribal culture that keeps them Fijian.

The recent political history of Fiji reads like a comic opera that begins sometime in the 1840s when a few village Fijians, irritated by something, burned down two houses belonging to the United States agent in Fiji, a Mr. Williams. An indignant Mr. Williams presented a claim for compensation to the chief of Fiji, Ratu Cakobau, the high chief of Bau, who had subdued most of the coastal tribes on the main islands and who considered himself king of Fiji. Mr. Williams wanted $20,000 for his two houses. I think Cakobau must have ignored him or perceived Williams as somewhat mad, and as Williams's indignation mounted so did his claim, until he was demanding $43,000 from Cakobau. In 1851 Williams's claim was reinforced by a visit from the United States warship the USS *Mary*.

Other things occupied Cakobau for the next few years, but the claim did not go away, and harassed by the U.S. government's insistence that he pay and the chaotic demands of European and American beachcombers-turned-residents, Cakobau applied to the English for help. Twenty-three years after the *Mary*'s visit, the English Crown agreed, among other things, to pay the American debt. In effect the United States lost possible control of Fiji for the cost of two wooden houses.

The great chiefs of Fiji ceded the land and sovereignty to Queen Victoria on October 10, 1874. In return they expected, and were granted, English protection of the land and culture of native peoples. When the English Crown returned Fiji to independence in 1970, it did not return sovereignty, the ownership and control of the nation, to the chiefs, but instead vested it in a parliamentary democracy. What was given was not returned.

As the numbers of Indians in Fiji increased, political power shifted into Indian hands. Democracy is not a Fijian idea. The rule and *mana* of chiefs are Fijian ideas, and have nothing to do with popular elections. Chiefs are supposed to rule the land, Fijians reason. It is their land, and it has always been ruled by high chiefs. As long as high chiefs are elected and the control of the government stays in Fijian hands, then democracy is tolerated, if not understood. It would have been more consistent with Fijian culture if the Queen had returned sovereignty over Fiji to the high chiefs, for it is the chief, the land, and the sea that compose the *vanua*, the place of the culture.

So there is hostility; sometimes it surfaces; sometimes it itches under the dark skin of both races. On the bus during the hot four-hour ride to Labasa from Nakadrudru, stopping interminably at nowhere to slowly load and unload passengers and goods, I watch Malé's subtle reactions to the Indians that board. He is impassively disdainful. His feelings never show on his face, but in the twitches of his muscles, in the clenching and unclenching of his fists, in his sometimes playful jabs at imaginary opponents. A bunch of young Indian boys jokes behind us. Malé's shoulders tense; he lights a cigarette and stares ahead of him. He does not consider fighting with Indians worth his time. They are too small and slender; they are not worthy opponents compared to the strong muscular Fijians of the *koro* and bush. Abstractly Malé hates Indians. In reality he is often compassionate and loving toward individuals, and invariably polite. But in no way does he consider them Fijians, and Fiji is Fiji. Is that right, Joana? Right?

Shopping in Labasa, Malé bargains over a 50-cent difference in a pair of shorts or a T-shirt. The prices of everything slide around like a greased pig. Nothing, except perhaps packaged food, sits still in price. Once I bought a $30-dollar dress for $8; another time the same storekeeper wouldn't budge past 25 cents. If we buy anything of substantial value, Malé tells me to wait outside because the storekeepers, seeing a "rich" European, will automatically up the price. Malé drags me from store to store in search of the best price. It doesn't matter whether I want to pay the difference and get going, preferring efficiency in the heat to the 50 cents, it is a matter of principled pride with him. Shopping is an ordeal for me. I see my face reflected in the shimmering shop windows. It is sweaty and gray.

In addition to waiting around in the store, I wait endlessly in the street while Malé stands and chats with friends and relatives. He never introduces me to his friends, and my impatience flowers. I try to act unconcerned, but it is clear that I am, for the time being, irrelevant.

Finally it is time for lunch, and we go into a cafe where wall-mounted fans try, hopelessly, to cut through the sluggish air. My dress sticks to the chair, and the undersides of my knees itch and burn. Malé drinks three quarts of blockbuster beer—one is enough to knock me over—and smokes an endless number of cigarettes. He looks at his plate of curried mutton, and I fidget.

I leave him over his beer and cigarettes to meet him in an hour at the market. I have to go to the police station to renew my visa, a monthly chore, and then will buy our staples at one of the two supermarkets in town. Away from Malé's imperceptible pace, my temper cools. At least I am on my way.

Cultural anthropology courses at UCLA offered me neat, tidy packages of customs wrapped in generalizations. Exotic cultures seemed like so much fun, mysterious and yet at the same time so straightforward. There were comments such as, in the spring the Lapps follow the melting snowline up to their summer pasturage. Anthropology courses hardly ever mentioned flies or boils or staph infections. They were strangely silent about toilets and garbage disposal. Culture was folklore, kinship, and rituals. There was no mention of heads of patriarchal clans chasing their daughters around in the middle of the night with a stick because they had forgotten to bring the water.

Now, having ventured into another culture, I am beginning to understand some of what all of those courses on culture left out. How long can you stand in the hot sun and say very little, if anything, to one another and be happy? Fijians very often seem to have nowhere to go. Stopping is more important than going; and they can stand around, each looking in a different direction, and say nothing, comforted by one another's presence, and not wish to insult anyone by seeming to be in a hurry. Americans become important by rushing; it shows that they are busy, that they have something to do. Fijians get embarrassed if they have to hurry. Fijians travel in clumps; they are not happy alone; they wither and become listless when forced to be alone. Most of the time when I am alone with Malé, he is silent, withdrawn, the light in his eyes turned inward. It is largely that we cannot joke and play in his language, something he loves, but it is more. I am still very much a stranger. I am always after him to spend time alone with me, to fulfill the image of the lovers. For him it's useless. What can we possibly do together?

But I am relieved to be alone and anonymous, and I push my way through the crowds on the street, happy to be moving. Groups of Fijians and Indians dam the flow, chatting and staring. A paper-thin old Indian woman hunches on the pavement smoking Fijian tobacco. A small pile of coins in front of her testifies to her poor day's take. There isn't much money to support beggars, yet the shops offer expensive silk embroidered saris, gold-coin jewelry, and $30,000-dollar trucks.

Almost everyone tries to look his or her best: Indian women in saris; young girls in sporty Western-style cotton skirts and tops; Fijian women regal in *jaba* and *sulu*, fitted tops over straight long skirts. The men dress in clean sports clothes, shirts and slacks, of varying expense and intensity of color. Some of the Fijian men still wear the tailored *sulu* with dress shirts. They look wonderful, important, like they are government workers or politicians. I wish Malé would wear his formal *sulu* in town as well as to church on Sundays, but he is enamored of Western clothes, prefers jeans and shirts, baseball caps and sunglasses.

Many of the Indian children are elaborately dressed, the babies decorated with plastic or gold bracelets and sometimes earrings. Their big round dark eyes smudged with kohl, they look like staring dolls.

Malé is at the market buying *yaqona*. It is his passion, and he shops carefully, going from stall to stall in search of the cleanest, strongest grog. It is in the presence of *yaqona*, cousins, and tobacco that he is truly happy. He is happy now for he has a kilo of *yaqona*, $12 worth, enough for an all-night session in the village, and the promise from me that we will buy a carton of beer to smuggle back into the *koro*.

We have left our purchases at the supermarket, the hardware store, and stacked with a vendor at the market; Malé is searching for a van, bargaining about the price, that can take us home to Nakadrudru. We have too many packages to go back on the bus; and I would rather spend $35 on the van than four hours on the bus.

Once in the van, a fairly new minibus driven by its young Indian owner, Malé leans back into the seat beside me, lights another cigarette, and reaches behind him into the carton that contains two pint bottles of Fiji Bitter. His old Suva style surfaces as he taps the cap of one bottle on the other to loosen the pressure and pops the top of the second with the first. He drinks the beer in the Fijian fashion, tilting it over his mouth and letting the stream flow into his throat without touching the bottle.

Malé is happy. He has his beer and the long ride to drink it. "Here, sweetie, have a *bilo*," and he pours me some in the tiny glass we bought to accompany the carton, puts his arm companionably around my shoulders, and settles in to enjoy the ride. He looks clean and relaxed; he's growing a beard, which accentuates the intensity of his eyes and makes him look older, more mature. He is dressed in blue—it's his favorite color—and the pale blue of his shirt compliments his dark skin. After a few bottles, Malé is content, singing songs for me, drawing my attention to animals along the way, yelling at friends passing in buses and lorries.

So we skitter and swerve along the winding gravel road between Labasa and Nakadrudru, passing heavily loaded cane trucks on their way to the mill and even more heavily burdened lumber trucks carrying immense logs from the bush to the sawmill at Malau. I'm thankful the beer makes me carefree because the driver, anxious to get back to Labasa before dark, is indifferent to life, caring only for time. Oh well! The horn is probably as reliable as the brake, and there is plenty of beer.

Vili is waiting for us with the boat in Nakadrudru, and we load the day's purchases on board and very slowly make our way home in the low-tide shallows of the river. By the time we reach the *koro*, the carton of beer, with the help of Vili, is finished, and it is dark.

FORBIDDEN RUM

When we were in Labasa I made the mistake of buying a bottle of rum in addition to the carton of beer we consumed on the way home. Wisely, the village meeting has ruled that no liquor be allowed on the island. But I had some fuzzy dreams of Malé and me, alone on Vedrala, sipping rum and watching the sun set over Yaqaga Island. These dreams bear no relation to reality, for I always find that Malé has other dreams, and more often than not they don't even include me. So the dream of the romantic drink together on Vedrala in the golden light of the setting sun has turned into a bitter argument because he wants me to give him the bottle to share with his cousins on Galoa after a night of drinking *yaqona*: "To wash down with the hot stuff."

I said no and gave my reasons, and Malé turned dark and inward, floating around in the night, not speaking, and avoiding my efforts to make of no a natural thing. I thought I would teach him a lesson, an impossible thing to do, and took a pillow and mat to sleep in Seine's unused *bure*. But I am no good at this kind of game playing, and, unhappy at missing him, I went back to our bed to sleep with him. He was gone, and I looked for him in the starlit sleeping village.

I found him in the pale light, sitting on a stone, as dense and rigid as the dead, holding a big kitchen knife hard in his clenched fist, staring into his own darkness. I sat next to him, rubbing his back, speaking softly and very little, trying to massage the misery out of him, and wondered if my life could end there under the flashing hurricane sky—the dream, the lover, the ghost.

The *tēvoro* lives in the *baka* tree. He is covered with hair the way the tree is covered with roots. He wanders through the village at night stealing souls. His victims cannot speak or move. I sat next to Malé—outside the young handsome dark prince, inside filled with the creeping mold of the *tēvoro*—and I said the Lord's Prayer, for him and me. I looked at the sky, and at the hurricane lights flashing their undecipherable message over the South Pacific sea.

Malé could not speak. I spoke little, for this man is unknown to me. I tried gently, very gently, to pry his hand away from the knife, but he only clenched it tighter. The stars were my comfort and calmed me. The heavens seemed so sure, so distant and uncaring, that they helped me keep my perspective on this glittering night smelling of death. I did not know whether Malé wanted to kill me or himself. It mattered little. My only thought was to sit there with him in the silence and find some means to radiate the calm of the heavens.

After a very long time he said, "Go to bed." With a final tentative caress to his unseeing eyes, I left quietly and went back into the eating house. I lay in the darkness and listened to the darkness. Malé finally came to bed and lay down beside me, not touching. The knife was not in his hands, so I felt a little more relaxed. When he was asleep, I looked for the knife and found it on the side of the bed. Very quietly I took it outside and hid it among the dishes and pots waiting to be washed and went back to a restless sleep, filled with foreboding. Was I willing to give myself to this man who could so easily turn dark and distant?

The morning brought some relief from the foreboding, but I was still uneasy. The morning also brought its burden of hurricane heat, and by midmorning I was in tears and ready to flee. A chance remark by Malé's father set me off, and I found myself throwing things into boxes and stuffing my suitcase in a frenzy. I looked at the *tabua*, the whale's tooth, yellow with age and heavy with ceremony, that Momo had given me to seal Malé's and my engagement. I had hung it over our bed against the beautifully patterned *tapa* cloth that was also part of the gift.

There was a time a few months ago when all I wanted in the whole world was the present of this gift, for much of my time had been given to whales, and it seemed an appropriate destiny to seal my future with the ancient ceremonial gift of the Fijian's—the *tabua*. I would leave the tooth and the *tapa* and all of the dreams behind. I would flee, admit that I had not the strength of mind to carry on this journey. When my things were together, I went down the stifling hot beach, burning my feet on the sand, to find Malé happily fishing with the children.

The sight of him, so free and uncaring, set me off again, and

I called him over roughly, "Come here, come here." He looked up in surprise. "I'm leaving. Take me to Nakadrudru," I said. "Take me to Nakadrudru. I'm leaving this place." He did not answer and threw the line once more into the shallow water. Furious, I went back into the house and started lugging my things into the boat. Kava, Malé's cousin, sat on the overturned punt and stayed silent, not offering to help. Ima watched me, wide-eyed, from the kitchen. After an unsuccessful attempt to put my heavy box into the boat moored offshore, struggling in the waist-deep water and getting more and more angry, I strode back to Malé. "Help me, at least help me put my things in the boat."

Finally he spoke. "Why are you leaving?" He looked at me from a great distance.

"Because last night you wanted to kill me, you wanted to kill me."

"No," he said, "I wanted to kill myself."

The hot still sky gave no answer. The warm dirty water lapped at the debris of the shoreline. The children kept on with their fishing. All was still, remote, uncaring.

"Not you. I wanted to kill myself."

For what? Because I did not give him a bottle of rum, because I said no. A great weight was in my heart, a great heaviness. I looked into his face, looked at the beautiful young face of a man in his prime, a man who wanted to kill himself over a bottle of rum.

I turned and went back into the house, not knowing what I wanted. Malé followed and looked at the chaotic results of my fear. He went to the wall and took the spring I had bought him, the spring that had held so much promise of self-improvement, of becoming strong and healthy, the promise of giving up smoking, of giving up all-night *yaqona* drinking, to train. To train. He took the spring and, saying nothing, slowly attached the long coiled bright cables to the framework.

"Here," he said, and threw it on the top of my things.

I remembered all of my moments like this—all of the times fear had driven me to snap the fragile threads of relationship—the pain of separation, the emptiness, the loss of love. It was a time to be very careful, to move slowly and with thought. Malé lay on the bed and played with a flashlight, saying nothing. I knew that if I asked him if he wanted me to leave, he would say it was up to me. It was up to me, and I looked at the impoverished house, at the woven bamboo walls, the *sulu* hanging from ropes, the disorder of my things on the floor, the bright curtain hanging limp in the heat, and kept silent.

After what seemed like a very long time, Malé got up from the bed and sat on the mat. He took a broken gas lamp from a corner and started taking it apart. Jone, his nephew, came and watched him as he performed the miracle of restoration. I looked at Malé, seemingly indifferent, repairing the gas lamp, and knew somehow that he cared. The improbable had occurred. Somehow we were constructing love out of an impossible beginning.

I lit a cigarette and turned and looked out of the window at the *baka* tree, at the home of the *tēvoro*. He had gone back inside his house.

I looked at Malé and said, "*Sa rauta, sa oti* (it's enough, it's finished)." The trouble was over.

Later in the day Malé came into the house and looked at my things, still in the boxes and suitcases. "*Isa*, Joana," he said. "*Isa*." *Isa*, the pain of regret, of longing, of remembrance. *Isalei*, you are leaving, I will remember you.

That night I decided I was too rigid, much too judgmental,

and no fun at all. I decided to get drunk on the bottle of forbidden rum, and when Malé was away cruising the *koro*, almost like a secret alcoholic, I found the bottle I had hidden under the bed and mixed myself a stiff drink with some water and fruit syrup. I lay on the bed, looking out at the night, and drank the rum by myself. The first drink made me feel much easier, so I mixed another, equally stiff, and by the time Malé came back, I was feeling no pain. Malé smelled the rum on me and asked if I was drunk. I said yes, and he wanted to know if it was good.

"Yes, it's very good. Do you want some?"

So goes resolution, from last night's insistence on not giving him the rum, to tonight, when I decided we should get drunk together. I mixed Malé a drink, and he drank it down, fast, almost in one swallow, which is his way of doing everything. All the way, no restraint, no thought of tomorrow. Okay, I thought, let's go for it.

Soon, in the way peculiar to the village, the room was filled with cousins, all of them drinking. How they knew we were drinking, I had no idea, but everything here is like that. It is just known somehow, and the forbidden liquor brought the boys clustering like flies. Sometime during the din, I fell asleep on the mat, lying next to Malé, a delighted host, who was carefully measuring out sticky sweet syrup, rum, and water and passing the common glass around. After a time the bottle was empty, and we dragged ourselves into bed and fell into a heavy drunken sleep.

Malé woke me in the middle of the night. "Joana, I'm sick, help me, I'm sick."

Good, I thought, it serves you right. Last night you wanted to kill yourself because I would not give you the rum, and tonight you're sick and think you are dying. Good, it's a lesson.

I turned on the flashlight and looked at Malé lying next to the open window. He was doubled up in pain and moaning, "I'm sick, Joana. I'm sick."

He was really sick, but I thought it was just the common sick of drunkenness and that as soon as he vomited it up he would feel better. I tried to get Malé to take some medicine the doctor had given him the week before for his constant stomachache, but he refused. Soon he was retching out the window, great gasping heaves that seemed as if his stomach would turn itself inside out.

I looked at the vomit by the light of the flashlight to see if there was blood in it. Only the sticky remains of the raspberry syrup pooled in the sand outside the window. After he finished vomiting, I thought he would be all right and told him to go to sleep. He lay there moaning, not sleeping, and I tried again to get him to take the medicine or drink some water to dilute the alcohol, but he still refused, all the while saying, "Help me, Joana, help me."

How could I help him if he refused the things I tried to do to help him? I found myself getting angry: stupid, stupid Malé, who had no idea of cause and effect as I understood it, wanting to kill himself, and now asking me if he would die.

"No, Malé, you won't die. It's just the rum."

"Please, please, Joana, help me," he cried. I broke into tears. I hate this place and these people who don't understand anything. And this good medicine I paid money for, you won't take.

Finally Malé swallowed a teaspoonful of the white chalky medicine and immediately vomited it up, the white smearing his face and dribbling down his chin. He looked so pitiful, so broken and hurt, that I realized that I knew nothing, could only offer the coldness of money and the burden of guilt that accompanies it. I realized that here in the night, with a suffering man I professed to love, I found only anger because he did not understand life the way I did, because I had wasted money on medicine he would not take. The useless medicine staining his face in the light of the flashlight seemed to say to me, you are wrong, Joana, you have no heart, only a wallet to trade for love.

Malé's cries woke his mother in Vili's *bure*, who came into the house carrying a kerosene lamp. I resented her presence. It seemed an intrusion. Why was she there in the night, getting in the way? Malé spoke to her in Fijian and then told me, "My uncle is coming to massage me." Again it seemed silly, unlikely. What good would a massage do when his gut was full of poison? I said nothing.

Uncle Luke and Aunty Una came into the house and sat on the mat. Malé's mother brought in the teakettle, and they all had a cup of tea, chatting pleasantly, acting as if nothing were wrong, while Malé moaned and writhed in pain.

Then Luke told Malé to lie on the mat on his back. Luke carefully poured coconut oil into his big cupped palm and began to massage Malé's belly. As I watched Luke work, I realized that I was watching a master. Something I had heard of for many years came alive before my eyes because of a bottle of rum.

Luke is a big man, fat for a Fijian, light colored, almost bald. I have always liked him, for from the first he called to me with a warmth that radiated kindness. He is not hard and muscular like most of the men in the *koro*, but round and jovial, laughing and generous. When I watched his hands on Malé's body, I knew that he really knew what he was doing. He was, indeed, a healer. I moved closer and forgot my anger and frustration, fascinated by his big strong hands pulling the poison out of Malé's belly, stroking and pulling the gut down, down, sparing no pain, for it was very painful, and Malé moaned and cried as Luke's hands massaged deep into his gut. While he was working, Luke talked and laughed, never mentioning the rum, but said that Malé had drunk too much *yaqona*.

"Yes," he could fix it this way. "Yes," he would teach me, and said, "Here, go this way, feel here where it's hot, feel here where the blood is throbbing. Yes, this way, this is the way."

Luke worked for a while and then let Malé rest, and then worked again, maybe for an hour, until Malé lay in a deep sleep on the mat and the heaving and twisting were gone. He put a washcloth soaked in cool water on Malé's belly then pulled a pillow next to the sleeping Malé. "Here," he said, "you lie down, next to him," and smiled a great warm generous smile that made my heart soft with love and respect.

"Thank you, thank you very much, *vina'a, vina'a va'a levu*, thank you. Will you teach me?" I asked.

"Yes, of course, I'll teach you." And Luke and Aunty Una left in the coolness of predawn.

I lay on the mat and thought about what I know, and what I don't know, and watched the sky lighten, listening to the roosters crowing, finally to fall asleep next to the sleeping Malé.

HEALING

A bright light is shining into my eyes. It is the sun. Scratch, scratch, scratch; Una is sweeping the mat with the *sāsā*, the broom made with the midrib of the coconut leaf. My head is heavy from last night's drinking and lack of sleep, so I lie with my eyes closed, holding back my irritation at being awakened. It is late, almost nine o'clock, and already hot. Malé is up and gone somewhere. I sit up on the mat and try to come awake.

"*Yadra*, Joana, *yadra* (good morning)," smiles Una cheerfully.

I manage a *yadra*, but wish people wouldn't talk to me before I am awake.

After a few minutes I stumble out of the eating house into the *vale ni sili* (bathing house), a small *bola* enclosure built next to the house on the high-tide line, which also serves as our impromptu toilet. Relieving my overfull bladder, I feel better and open my eyes some more. No one is in sight.

When I go back into the eating house, Una has put away the pillows, finished sweeping the mat, and set out a cup of tea and a dish of pancakes. Drinking the tea and passing on the pancakes, I ask, "*Malé e vei* (where is Malé)?"—a question I ask often enough to have learned how to say it in Fijian.

Una, not fooled by my halting Fijian, answers in English, "He has gone to see Uncle Luke and will be right back."

Knowing better, for Malé's "right back" often stretches into hours, I finish my tea, splash some cool water on my face from a bucket in the lean-to kitchen, straighten the *sulu* I slept in, and go out in search of Malé.

He is sitting in the shade with Uncle Veresa and Uncle Jo. Uncle Veresa has his *sele* propped on his bare foot and is sharpening it with a file. Malé and Uncle Jo are smoking home-rolled Fijian tobacco and telling some kind of a fishing story. I know it's a fishing story because I recognize the names of fish and there is much measuring of size. It must have been a big fish, for Malé has his arm stretched full out and his other hand crossed on his shoulder. He looks at me and continues talking. I wait.

"Joana, no way you could drink the medicine. I like throw out, it's really bitter," Malé tells me after a break in the Fijian. "You go try. See if you can drink." Malé is talking about some Fijian medicine that Uncle Luke has given him for his sickness of last night. "Uncle Luke there," says Malé, pointing to the other side of Luke's small *bure*, where he and Aunty Una have their cooking house. "Go!"

The thought of drinking bitter Fijian medicine and throwing up is not appealing, but neither is sitting in the heat listening to incomprehensible fishing stories. Besides I am interested in learning more about what Uncle Luke knows about medicine and massage, and want to prove to Malé that I am not a total wimp. If he can drink the medicine, so can I, so I tell Malé, "Okay, I try," and go off to look for Uncle Luke.

I find Aunty Una sitting cross-legged on the sand outside the cooking house, cutting breadfruit and putting the quartered pieces in a big pot to boil. She stokes the fire with some dry coconut husks and smiles, "*Yadra*, Joana. Luke there, go inside."

Uncle Luke is sitting on the mat near the doorway, drinking tea out of a tin bowl. A sagging food chest leans in the corner behind him, its screen protection against the flies long gone. Opposite hangs an immense bunch of dry *voivoi* (pandanus) leaves. Rolled up beneath the leaves is an unfinished mat. A few pillows stacked in a far corner, three suitcases stuffed with clothes, a basket of soiled clothes, a tiny transistor radio, and a hank of *magimagi* hung from a corner post complete the furnishings of this tiny house. The only other thing that catches my eye is a tin bowl filled with a scummy green liquid—the medicine.

I refuse another bowl of tea—the sweat is rolling down my face already—and I listen as Uncle Luke tells me that he went into the bush just after dawn to get the medicine, a root from a small lemonlike tree that he washed and scraped, infusing the scrapings in a clean cloth with cool water, just like the *yaqona*. Luke smiles, looks gently at me, and asks if I want to try the medicine. "It's good for you, it cleans the stomach."

Luke carefully rolls some Fijian leaf tobacco in a dry *susuna* leaf, offers it to me, and we smoke together in silence, listening to the laughter from outside as the fishing story continues. After delaying as long as possible with the smoke, I gulp down the medicine, expecting the worst. It is cold and slightly slimy, but nowhere as bad as Malé would have it. Aunty Una comes in smiling. Her eyes are deep-set and full of warmth. She has the kindest aura of anyone I have ever met, and she is skilled in medicine and massage and is a master weaver. Luke with his great strong hands massages men; Una with her gentler touch takes care of women. They seem to be available to anyone who comes asking for help, but do not seek out patients. It's something for the family.

Luke tells me that he learned his medicine from his father, who learned it from an old woman. Aunty Una learned from her mother. There is nothing secretive about either of them; they are willing to teach and share. They tell me that there is men's medicine and women's medicine, and some common to both. Coughs, headaches, body aches, wounds, sprains, and breaks are treated commonly. Women's medicine is for childbirth, to clean the womb, and for infections of the reproductive system. They explain a curious medicine to take if you are pregnant and nobody knows. There is medicine if you have a headache at sunrise and sunset, but not in between, and medicine to take if your eyes won't open in the morning. Leaves, roots, bark, and the scrapings of stems are used. Some are compounded of several different plants, some used alone. All are specific. Some medicines are drunk in an infusion, some chewed, some used as a poultice.

Una explains a poultice for drawing boils; it is compounded of three different leaves pounded with a spoonful of sugar and some shaved laundry soap. Laundry soap and sugar elude me, but I am willing to try anything, for I am plagued with boils. Una promises to bring me some the next morning. By now it is much too hot to go back into the bush.

Luke knows a little English, enough so we can stumble along. Una knows only a few words, but her eyes make up for the lack of language. Being originally from different villages—Uncle Luke comes from up the coast in Macuata—they argue lightly about how to prepare a medicine. Gathering and preparing medicine takes time and effort; it's not as easy as popping a pill. As I watch them disagree, I remember Luke's hands last night, glistening with coconut oil, pushing the pain out of Malé's gut. Their craft is more elusive than this beginning conversation can possibly reveal. I come back to the medicine that you take if no one knows you're pregnant. "Is it to get rid of the baby?"

"*Maqa* (no), not like that. It's like if you go to the water tank to get water and you have the baby inside, and no one knows."

"Oh," I answer, more mystified than ever.

I ask if they have to pray or say or think certain things when they gather or prepare or give the medicine. No. Prayer, Christian prayer, they say, is separate from the medicine. I remember that my father-in-law went to the other side of the *koro* last week to pray for Uncle Paula's mother, who was sick. The medicine is just that: you gather, prepare, and give it without ritual.

Malé is coughing outside. I can track him around the village by his cough: dry and scratchy and unmistakable. It seems as if he has been sick for weeks now—sometimes coughing and feverish, sleeping all day; at other times doubled over with intense stomach pains that come soon after he eats. I worry about him, for he is losing weight and seems listless, yet he refuses to stop smoking or drinking *yaqona* all night, which is always accompanied by constant smoking and little eating. He responds to my health lectures with blank silence, does not understand cause and effect the same way I do.

Fijians pride themselves on the ability to endure (*vosota*). Life and health are given by God and taken by God. Life is perilous; the most important thing to do is to enjoy it day by day and not cloud it with too much reasoning. Why? is an imcomprehensible question. I ask why someone is sick, or what kind of sick. I get the answer: I don't know; he's sick. Why did she die? I don't know; she died.

Although there is a government health center nearby, at Nakadrudru, which dispenses free drugs and provides a paramedic to do examinations for twenty cents, few of the people in the village go there when they are sick. They are afraid of the "doctor," afraid of his ever-present needle; and often they lack the twenty cents or the money for the benzine.

A few weeks ago Malé stepped on a nail that went a good half inch into the bottom of his foot. He treated the wound by pounding it hard with a hammer and then burning it with a match. The next day I asked him to go to the health center for a tetanus shot, and he refused. He was afraid of the needle. I managed to talk him into going by promising to buy some *yaqona* if we went to Nakadrudru, and then cornered a respected cousin to assist me in getting Malé in to see the nurse. I watched while he received his dreaded tetanus shot. "But, Joana, it didn't hurt."

"I told you it wouldn't hurt. Now will you believe me when I tell you something?"

Malé grinned. It was clear that he wouldn't.

I ask Luke about Malé's recurring stomachache and cough. He suggests that Malé get up early in the morning and bathe in the sea when it is cool and then come back and drink some more lemon-root medicine. I promise to bring Malé back tomorrow, but am not hopeful.

Luke, Una, and I share one last home-rolled smoke. I gather up my notebook, pen, and tobacco and thank them for the *tala-noa* (conversation) and the smoke, slip on my flip-flops in the doorway, and go outside. It is as hot as a sauna, and Malé has vanished.

⁂

Tonight is soft. There is a gentle wind, cool enough to make me want to wear a shirt. Malé is drinking *yaqona* with the men, a floating conversation underlaid with the melodies of the guitar and the ukulele. From the *vatu nu loa* I hear Nei working the *voivoi* as she prepares the leaves to make mats, a soft scraping sound. The *koro* is restful and still. Uncle Jo's wife, Mada, just came back from fishing on the reef with a basket full of huge land crabs, claws neatly tied with mangrove string. She hunted them in their hiding places in the mangroves. She will take them to Lekutu tomorrow to sell them. Everything seems resolved and at rest.

Is it the *yaqona* that brings this feeling, this calm repose that frees me from needing anything other than this night, these stars, and the soft singing? I look at the men sitting with such comfort on the grass: strong bodies, muscular arms and chests, the seductive careless way they wrap their *sulu* when they get up, the pervasive courtesy of their talk and motion. A timeless grace drifts here, an eternal dream time of people comfortable with where they are and with each other. I have no idea what *yaqona* does to the mind or to the body; its chemistry eludes me. But it confers a singular and unique calm. The *tanoa* reflects the stars overhead, a miniature universe.

This afternoon I let my worry about Malé surface. Perhaps his sickness was because of me and the demands of my Western ways; perhaps inside he could not accept them, but was caught in the desire of his father that we marry. I went to Momo to talk about it. With our broken speaking translated by Una, I conveyed my fear. Did Malé unconsciously want to die to escape this bind? Malé had told me that the *tēvoro* had told him that he would die on March 13. That was two weeks away.

Fijians believe that if something is wrong, if an unknown sin has been committed against the spirit of an ancestor—the *vū*—someone in the family will die. Was Malé slowly fading because something was wrong?

Momo said I must bring him some *yaqona*, and we would hold a *soro*, a cleansing and clearing of whatever might have been done, whatever might be wrong that we didn't know. This evening, after dinner, I brought the *yaqona* to Momo. We sat in the grass outside the eating house under a star-sprinkled sky. I made a simple speech, translated by Malé. Presenting the bundle of *qona* to Momo, I asked God to restore Malé's health, asked forgiveness for whatever wrong might have been done. I felt the calm of the Fijian way, the acceptance of the wisdom of that way. This was no analytical questing into the whys and wherefores of the problem. Much simpler than that, it was the acknowledgment of the ways of the universe and man, and the recognition that we do not and cannot know; we can only believe.

Malé felt immediate relief, and his temper improved; the depression and lassitude disappeared. We sat with the family and drank the *yaqona*, joking and laughing. We were doing what had to be done to solve the problem. It used to be that the *tabua* was presented in a *soro*. Now, in the absence and scarcity of the *tabua*, it can be *yaqona*, or if no *yaqona* is available, even two bars of laundry soap. A cleansing.

Scrape, scrape, scrape; the *voivoi* is softened in preparation for mat making. The *bilo* is passed, and the *yaqona* is drunk. The incoming tide flows over the shore a few yards from where we sit.

The kerosene lamp burns steady and bright. All is forgiven; all is harmonious in the universe.

NESTING

The heat was getting to me. I was becoming a tyrant, a cost accountant of powdered milk, a detective of two potatoes. One morning last week I was awakened out of a heat-drugged sleep by the crying voice of Sala, Kava's wife and a cousin of Malé's, who was sitting on our mat weeping because a neighbor had accused her of stealing some of our cocoa. Dismayed at her early-morning intrusion into my private life, I shooed her out of our *bure*, rudely forgetting that it was the family's *bure* and that she was a member of the family; and, reinforcing the accusation, I looked into the cocoa can to determine if there was less there than I remembered. Malé said nothing, but looked at me as if I were indeed from another planet. A few minutes later he responded to a familiar laughing voice outside and slid out, leaving me with my lonely frustrated jealousy.

We have finally moved into Seine's vacant *bure*, something I have wanted to do since she left the *koro* to visit her family in Suva. Malé could not see any reason for us to move out of the eating house, for he is comfortable with the 24-hour presence of his family and sleeping in the same room with his sisters and nephews. I was not, and was slowly cracking under the strain of incomprehensible foreign voices, flies, and boils that wouldn't heal.

After Malé left that morning, I wandered around the village looking for some place where I could be alone and write. I needed the time with myself and the comfort of my own language, even if it was only the monologue of writing.

The constant swirl around me of fast-spoken Fijian, which I still couldn't understand, except for a few words and phrases, kept me from even thinking in my own language. I was used to privacy; my moodiness required time to sort out my emotions, to pick my way through my thoughts. Malé and I were never alone for more than a few minutes. Too many times when I was trying to explain myself to Malé, a rustle at the doorway would interrupt, and Malé would turn away from the difficulties of trying to understand me and give his attention to some cousin or friend waiting outside to play. If I went to sit on the beach or to wander in the bush to be alone, inevitably someone would come and join me and confuse the flow of my thought with simplicities that left me feeling as if I were a child.

I was drawn to Malé; his energy and straightforward ways appealed to me. If only I could get his undivided attention once in a while, our way would become easier. But without a house of our own, without privacy, it was hopeless.

Later that morning I encountered Saraqia, the chief's son, who spoke English, who had once told me that he would be pleased to help me if I ever needed anything. I had talked with Saraqia one day after church, fascinated with his blazing sermon, which seemed inspired more by the devil than by a kind and merciful God. Unaware that Malé and Saraqia were not the best of friends, I asked him if he knew of a place in the village where I might be alone and write.

He presented me with the dubious loan of an old storeroom, which had once been the village cooperative store, tin roofed and in a place guaranteed not to catch the slightest breeze. It was awful, but I had gotten myself into it and couldn't gracefully refuse his generosity. Saraqia got the key and assured me that the storeroom was his and that I could use it as long as I wanted to.

Not letting me slide away until a better tomorrow, Saraqia immediately cleaned the place with great energy, raising clouds of dust and the heavy smell of rats. I poked around in the mess, more interested in the contents of old cartons than in accomplishing something, found a big wooden box to sit on, and set my typewriter on the splintery counter. I neatly piled my paper to one side and sat down looking thoughtful.

Saraqia left, quite pleased, and I sat pretending to write. I was not there a half hour when I heard something hard falling on the roof, followed immediately by another something hard. Malé was throwing rocks at the building.

The problem was simple. If I used the storeroom, which belonged to the whole village, not to Saraqia, without first getting the permission of the village meeting, then everyone would be jealous and angry. It mattered little that Saraqia was the son of the chief and might someday be chief. It was not his storeroom, but the village's, and I was wrong to use it.

I was glad for an excuse to leave. There was no way I would have produced anything from that tin-roofed chamber except sweat-stained pages of whining, so I gladly took advantage of Malé's reasoning and sweetly told him that if he didn't want me to do it, of course, I would not use the building. He must have been pretty surprised, for my temperament these last months has been to fight with him over how many cigarettes he smoked in an hour, or who has been stealing the instant coffee. I have, in truth, not been the perfect companion, and was somewhat relieved to find myself agreeing with Malé for once, whatever my motive.

Somehow Malé finally got the point, and that afternoon, when I returned to the eating house to store my typewriter under the bed, he told me that I could use Seine's *bure* if I wanted a place to write. I had asked him if we could move into Seine's *bure* two months earlier, for she had left shortly before Christmas, and the *bure*, which is next door to the family's lean-to kitchen, has been empty ever since. All it needed was a cleaning out.

Malé went off fishing after lunch, and with my last bit of strength, I hauled the dusty mats out of the *bure* and looked in dismay at the cobwebs and the rat and chicken litter and smelled the discouraging scent of the unused grass house. Una came in cheerfully and began to haul and pull, and soon her cousin Marselina appeared; and both of them, with more strength and assurance than I, and also with more knowledge, set about to clean the *bure*.

They pulled the rest of the mats out and laid them in the sun to dry and air; then they intrepidly gathered up great armloads of dried coconut leaves, the *sāsā* that cushions the mats, and strewed them in the sun to dry. We swept the sand floor of its accumulated treasures—a rusty razor blade, fragments of tea cartons, cigarette filters (there are never any butts; they are smoked down to the end)—removed the assorted decorations stuck in the *bola*, and cleaned out the corner hearth of ancient ashes. In less than three hours Una and Marselina managed to change the musty *bure* into a possible home. What was offered as my studio was going to be transformed into our house. It was my surprise for Malé.

We covered a worn low table with a bright *sulu*, propped my books and notebooks on a shelf, covered another old board with a red-and-blue *sulu* depicting a blue ocean, yellow sky, and red foreground, announcing in blue letters that "Fiji Is the Way the World Should Be." On it I arranged my precious hoard of instant coffee, cocoa, powdered milk, eggs, and oatmeal in what was to be the cooking corner. With the addition of the Primus stove, the foam mattress, the mosquito net over the bed, the "designer" fitted sheet, our clothes, and two jars of flowers, I was home. I happily awaited Malé's return. When Malé came back from fishing, he walked into the *bure*, got a cigarette, and left without a word. I burst into tears.

Hurricane heat, an alien culture, boils on my head, and, indeed, dear Mother, what was I doing here?

The next weeks brought more tears and hourly resolutions to pack up and leave. That only thing that kept me from leaving was the long bus ride to Labasa in the heat, the realization that I was a sniveling coward unable to cope with the reality of my fantasies, and the disconcerting knowledge that I was falling in love with Malé and that he was slowly changing under my nagging into a sometimes approximation of the hero I dreamed him to be.

I, however, was definitely not heroine material. We fought and fought. I was disdainful of his culture, confusing my personal unhappiness with flaws in the Fijian way. Why did he drink *yaqona* all night and sleep all day? Why didn't we walk hand in hand in the sunset? Why did he prefer the company of his joking and laughing hardy family to my whimpering and penny-pinching? Didn't he understand that I was spending money on him? Didn't he understand Europeans were different? Why didn't he kiss and hug me like in the movies?

Ever since I can remember I have equated love with gentle caresses, and the tears of the lover with the "there, there, everything will be all right" response. When I cried, Malé just looked at me. After what seemed like a light-year of tears, I began to look around.

Fijian babies are treated roughly. As soon as they show any strength, that strength is encouraged. They are slapped on the head in play and pinched on the bottom in affection. Their bodies are prodded and poked and their private parts tickled and twisted. They are raised to withstand the rigors of a hard life and to endure, and their bodies and psyches are prepared for it. True, they are held and cuddled, especially the infants, but as soon as they are old enough, they are treated in a rough-and-tumble way. Adult Fijians never kiss one another; it's not their way. When family members meet, they sniff each other's faces. Men and women never display open affection, are hardly ever seen together. For the longest time I had no idea who was married to whom, for I rarely saw a man and wife together. Men work and play together, and women work and play together. My attempts to be with Malé often embarrassed him. I had succeeded in establishing the convention that I would sit next to Malé while he drank *yaqona*. It was the only way I could spend any time with him, for he drank *yaqona* nightly, but although he sat next to me for hours, he hardly spoke to me, all the while joking with everybody else. When I asked him why he didn't joke with me, he said, "Because it makes you mad." And it was true that a hard slap on the forehead or a punch in the ribs was not my idea of love.

Malé had not been raised to be the perfect lover as I understood it. And here in my tropical dream life I was more demanding of the perfect lover than I would have been at home, where I had much more to reinforce my self-esteem and rationality. Malé was simply being Fijian, not a monster, and slowly the light began to dawn. Yes, he would learn my ways; yes, he wanted to learn my ways, but to expect him to learn after a nonstop nagging session, where I only repeated how sick or weak or unhappy I was, was asking the impossible. Fijian women are loved because they are strong and can contribute to the household. I would have to learn too—learn how to teach, learn to see myself as a mysterious bundle of incomprehensible desires. I would also have to learn how to accept Malé's demonstrations of affection, demonstrations that often left me feeling assaulted or humiliated rather than loved. I would have to learn to understand love in an entirely new way. Creating the dream world was not going to be as easy as creating the world.

Ease up, Joana, you came here to learn, not to reform. If you want to learn how to live joyfully in the present, you cannot accomplish it by keeping track of spoonfuls of instant coffee.

Now, blessedly, the heat seems to have broken. For the last two days the trade winds have blown fresh air into the *bure*. I have my table and typewriter, mosquito net, and my jars of flowers. And, thank goodness, two days ago Ben, the young man from the Peace Corps who is organizing the cooperative fish market in Lekutu, lent me a copy of *Jitterbug Perfume*.

Thank you, Ben, for the book, and thank you, Tom Robbins, for the refresher course in human history. Life seen through the eyes of the tragedian will always be a tragedy, and life seen through the eyes of the perfectionist will always be imperfect. Life seen through the eyes of the humorist, well, it just might be funny. And, besides, my boils have finally broken, spewing out their disgusting contents of pus and blood, and I am beginning to feel like I just might live.

RESURRECTION

The village huddles under its burden of noonday heat. The *bure* squat like shaggy beasts panting in the fierce sun. Only the chickens stir, mouths agape, scratching and pecking along the shoreline, digging in the dirt with their peculiar chicken motions that remind me of those little wooden toys: pull the string and the chicken bends and pecks; release it and she stands expectant. Cousin Toa's newborn white piglets are asleep in the shade, silvery flanks heaving in piglet dreams. The goats are resting in the nonexistent shade of the *koa* bushes they browse so relentlessly. Even the crabs are cooling their claws in their sand caverns.

From inside the church I hear hammering, a readying of the new church in time, we hope, to pronounce the resurrection. On the fires the pots boil the daily ration of cassava and breadfruit. Women's voices rise and stir, laugh and scold. The great feathery *vaivai* trees have shed their glory of flame blossoms and now, still and green, wait for the fluttering of the *cagi bula*, the refreshing wind that brings relief to the land.

In the summer the winds blow from the north, hot sticky equatorial winds that offer no relief from the hurricane heat. When summer turns into fall, starting in March and gaining impetus in April, the winds shift to the southeast, trade winds—the life-giving cool winds. We are in the interlude between the two. Some days are blessed with the trades; then a morning such as this comes, with the hot heavy north wind barely stirring. I long for the hurricane season to be over so that I might rise from sleep refreshed and ready, not heavy and irritable.

Last night it rained hard, very hard, a drenching frightening rain accompanied by great flashes of lightning nearby and the constant roaring and rolling of the heavens. Inside the *bure* it was like a steam bath: windows blocked by sheets of corrugated iron as protection from the rain, no air inside, the kerosene lamps adding their burden of heat. We rolled and smoked cigarettes and listened to tape cassettes of Hawaiian music to overcome the lassitude, almost naked in the near darkness, sweat rolling off our bodies, teasing and joking and poking to retain humor and maintain happiness.

"I want to be happy every day," says Malé, and flips the cassette. "Music makes me happy," and he looks at me expectantly. "Right? Right!"

Now it is midmorning, and Malé has gone to fulfill his village obligations to the new church. The men are cementing and plastering the walls, painting them a soft green that I never would have picked for a church interior, yet which seems surprisingly appropriate. *Bula* (life). We thank God for the life.

Life in this village is a continuous ceremony. A soft spirituality pervades. Every meal is graced by prayer. Even the social drinking of *yaqona* is preceded and followed by the appropriate ceremony and thanksgiving. Life is appreciated; hard and rigorous, it is relieved by laughter, an offering of humor and manners. Personal moodiness is disdained. One is expected to grease the social life with polite good humor. *Yadra* (good morning). *Mai'ana* (come and eat). *Moce, roaroa* (good night, tomorrow). It is a dance of manners, assured, customary; it grants order and peace to life.

The church stands in the center of the *koro*, higher and more imposing than any private dwelling. It centers and anchors the social life and provides the people with all of the philosophy they need. The *lali*, the big hollowed-out log that serves as the village drum, once called men to war; now it calls them to worship. Fijians are proud of their religion, primarily Methodist, and thankful to the early missionaries for guiding the culture away from cannibalism. They speak of that past, not with shame, but with an open gratitude that it is the past. It was not a pleasant history, and more often than not it was the weak—old people, women, and children—who were preyed upon. And it was not so very long ago.

I walk over to the new church to see what Malé is doing. He is high on a scaffold built against the end wall, dangling his feet over the side, smoking and joking with Qare; Toa, his closest cousin; and some of the young boys. They all call good morning and beg cigarettes from me.

"Joana, look," says Malé, and he grabs a tray loaded with cement from Toa, who playfully punches his shoulder. "Look, Joana! I know how," and Malé slathers some cement on the wall, showing off his new-found skill. I joke with Toa, saying Malé works only when he has an audience, hand out two cigarettes, and watch Malé working. The hard morning sun is softened by the new stained-glass window; above him pastel light molds Malé's muscles, transforming him into a Rembrandt image. The light joking moment is suffused with holiness.

The new church has been more than twenty years in the building. It is a massive achievement for this village of twenty-six families. It has taken so long because all of the materials have had to be bought dollar for dollar. The labor has all been free. The chief's brother Uqe is the principal architect; it was his idea and his plan, and over the years, the small *soli* (gifts) to the church have bought lumber, blocks, roofing, cement, windows, and doors to construct this village showplace.

Built of concrete block, plastered inside and out, with a corrugated roof, the church is a big cool structure with a high wooden ceiling and a row of doors opening along both sides. A few years ago the village was visited by an Australian woman who came as a tourist and took a liking to the people. She has helped them out over the years, providing some of the money for the construction of the cement water-catchment tank and for making two stained-glass windows for the new church. There may be only five or six churches in all of Fiji that have stained-glass windows; and for this small impoverished village to have such a special and exotic acquisition is a source of much pride. For the last two months the whole village has been occupied in finishing the church and getting ready for the great opening and the visit of the Australian woman.

Each woman in the village must make two mats to present to the visitors; the village meeting house must be rebuilt; and a big thatched *bure* must be constructed for the Australians. Then the village has to be spruced up, all of the weeds and grass around the compounds cut by hand with machetes, and a great feast prepared for those who will come for the opening. All of that represents a massive amount of labor for a small village already burdened with the daily labor of growing and harvesting food, fishing, cutting and carrying firewood, getting water, washing clothes, cooking food, washing dishes, making household mats, and cleaning compounds.

For the last two weeks Malé's mother and sisters have been cutting and preparing *voivoi* to make mats. Late into the night I hear the scraping of the *voivoi*. Passing the *bure*, I see inside, in the dim light of a kerosene lamp, the women bent over their mats. While the women sit alone in the *bure* weaving, the men gather outside and drink *yaqona*, laughing and joking.

The mats are exquisite works of art, elaborate and fine. A bright varicolored wool fringe, echoing an ancient pattern of brilliant red feathers, decorates the edges. Bed mats are woven with intricate black-and-white designs and heavy rows of wool fringe, fashioned to hang down the side. Floor mats are a bigger, coarser weave and have intricate traditional edges. Last month Monomono needed a big floor mat and, working as a schoolteacher, was unable to find time to make her own mat. She commissioned a number of women from the village and, in return for cooking for them while they wove, was given a 12-foot-by-18-foot mat, completed in two days of sociable weaving by eight women.

Usually two or three women of a household will go into the bush and spend an afternoon cutting *voivoi* leaves from trees planted in the family plantation. Returning with their heavy brilliant green burdens of saw-toothed leaves, they plunk them in the sun and spend one or two afternoons stripping the fine sharp edges off the leaves and cutting off the midribs. Then the *voivoi* is laid out in the sun to dry, decorating the compound with a geometric floor of bright green turning to soft tan. When the leaves are dry, they are softened by pulling them against an upright post and rolled into big wheels. When it is time to begin the mat, the long leaves are cut into even strips with a simple kitchen knife, and the weaving begins.

A small mat would take about two days of uninterrupted labor, much more if it is decorated with wool fringe. Uninterrupted labor is impossible, so the mat making is squeezed in between other tasks; and in the late hours, when the children are asleep and the men happily congregated around their *yaqona* bowl, the women weave and weave. Passing the *bure*, I look at the women bent over mats and wonder what thoughts occupy them in these soft hours of silence and solitude.

These patterns of mat and *tapa*, the basketwork interior of the *bure*, the gentle colors of tan and brown, the intricacy and regularity of the weaving, all speak of Fiji. It is as if these patterns give life its order and calm. Everywhere I look I see the patient work of hands taught by generations of repetition to perfect these simple and sustaining textures. The people are harmonious with the colors that surround them, dark brown gleaming skin enhanced by the pale brown walls of the *bure*, the golden sheen of mats, the intricate rich black-and-brown-and-white patterns of the *tapa*.

⁂ ⁂ ⁂

Nei presented me with a mat yesterday, a rather large plain mat, perfectly woven to fit our small sleeping house on Vedrala. The house is finished; it waits there for our residency. We haven't been to Vedrala for weeks and weeks; it seems there is always something more important to do that takes up our time. I think Malé is afraid of living there alone, afraid of *tēvoro*, and not sure what we would actually do if we were alone together. So the dream waits; and while it waits, I am learning to be patient, to take each day as it is given, to learn to say yes when asked by Malé, "Please, please, Joana, may I go fishing?" or "Please, please, can I drink *yaqona*?"

At first I tried to control our life together, to plan it according to my desires and whims. But I found that every time I said, "No, no, I don't want to do that today," or "No, stay with me here and go fishing tomorrow," it would backfire. What could have been simple turned into a morass of hurt feelings and confusion.

The simple truth is that in spite of my extravagant education in "primitive culture," I haven't the slightest idea what lies behind the unfolding of daily life in this village. If the tide is right or the time of the month is appropriate, that is when the fishing must be done, not when I am in the mood to be alone and can spare Malé. So after a series of disasters, which more often than not culminated in a tense withdrawn Malé and a lonely and hysterical Joana, I told myself that I would try to say yes (*īo*), yes, do it, and later find out the reason. Since turning that corner of the mind, I am finding that life is easier and happier, and that Malé tries to be the lover-hero I would wish him to be.

I am entranced by his innocent goodness, by his deep love of his family and his growing affection for me as I show him that I, too, am kind and care for those he loves. I have never met a man who could so easily forgive and forget, who did not gather his grievances and use them as weapons. *Sa rauta* (it's enough); *sa oti* (it's finished). And it was finished. The harsh words, the whining and crying, would wash away as easily as the tide, leaving a clean beach upon which to write the next day's events. Perceiving this, I find that I really do love this man, with a love that surpasses my desire for him or my appreciation of his beauty.

When I first came here, I was preoccupied by the dream life, by making the travel poster into reality. But the romantic vision of lovers gazing into each other's eyes, of touching sparkling glasses of chilled wine with the sun setting behind the seductive curve of bending palms has nothing to do with the realities of subsistence life in a Fijian village, has no relationship to living in the confines and security of a large extended family. The dream image then must be seen for what it is, a photograph designed to sell airplane tickets. Yet I am grateful for the poster, for without it I might never have ventured into this strange land.

I remember a dream I once had of a blue egg. I was in love then, and I searched for the blue egg to give to my lover. I looked in stores and under the fat rumps of sitting hens for the blue egg. I felt somehow that if I found that perfect blue egg and gave it as a gift, life would become good for me. I fantasized how I would wrap and present it—the blue just the right color; the egg, the perfect egg. I found cheap substitutes: Italian alabaster eggs; painted Easter eggs; and once, many years later, I found a sculptured marble egg with the chick folded inside, ready to come out, the chick sleeping in the egg, ready for birth. I gave that egg, in a satin pouch, to another man I fancied I loved.

On Easter morning I boiled eight dozen eggs and transformed them with food coloring into glowing jewel eggs of lavender, deep blue, and emerald green. The family sat around me, charmed by this latest European trick, for no one had ever before seen a colored Easter egg. When we had painted the eggs, we put them into little woven coconut-leaf baskets, produced almost miraculously the day before by Nei, Seine, and Aunty Amele. We added a slice of watermelon; some bright flowers, hibiscus and frangipani; two pieces of hard candy; and half a peeled orange and called the children of the *koro* to come get their Easter baskets.

The baskets looked brave and bright waiting for the children, arranged on a tablecloth on the mat in the *vatu nu loa*. Toa, Qare,

Seine, Vili, and Malé's sisters gathered, laughing, intrigued and expectant. Something new here, this giving of the symbol of the promise. Then the children all came, bewildered by why they had been summoned, and stood in a quiet orderly line while we handed out the baskets. Well mannered and confused, they stood around holding their Easter baskets, not having the slightest idea of why they were being given this gift, not having the slightest idea of the meaning of the colored egg.

Seine, Vaseva, and Ima shepherded them into an orderly group, and I took a picture of the village children holding up their baskets, the smallest ones barely able to stand, and then they all went home to eat their eggs—for some the only eggs they would eat this year, or next year—and we cleaned up the leavings and smoked the last of the tobacco.

Inside the old church it is hot and humid. The pastor is preaching, and his words sound angry. The short man with the hump is important today, for he has the long stick with which to hit the children if they misbehave, and he is serious about his job, for it is important to keep order in the church. While the pastor preaches on the meaning of the resurrection, the humpbacked man walks up and down the center aisle, hitting the children. Outside it is raining. I see the coconut palms misty in the gray light, dripping rain. The grass has the clean color of rain-washed earth. Inside, the pastor talks on and on. I hear no promise in his voice, and the man hits the children. The only promise I see is in the green earth's receiving this heavy intense rain.

There is no prayer in my heart. I am distracted by the scowling faces, by the bowed heads, by the lack of joy and color. I feel the weight of Christianity, the heaviness of heads always bowed, bent under the burden of prayer. The singing is sweet and expert, but seems without feeling—only the orderly almost-perfect rendering of the chromatic scale; the pure blending of alto, tenor, and bass; the sopranos reaching.

The church is old and bare; some of the pews are broken, and in the last row the back is off one, forcing the men to sit bowed forward. It is dark inside; the gray sky gives no light. I smell the odor of people sweating in the damp hot air. The man with the stick, more important than ever, closes the doors in the back of the church. I ask him to open them, for it is stifling hot, and he reluctantly opens one door a bit. I want to take the stick out of his hand and beat him on his hump. Instead I look out the window at the rain and meditate on the promise of the resurrection.

CHILDREN

Morning light splashes gold on the leaves of the ginger planted in a border around the *bure*, enchants the intense red leaves of the croton, illuminates the grass. I turn and look at Malé's head in profile, deep asleep, his eyelashes folded over his uncertainties, his breathing regular and assured. I lie on my back on the mattress and look at the world through the oval archway of the *bure* door, a world perfect and still, blessed with a cool breeze. There is no sound except the faint chittering of the mynas. Then I see Toa walking by, carrying his *sele*, on his way to the bush. A moment later his wife, Sala, passes, a blur of red and blue light, on her way to the water tank. I lie on my back looking out the doorway, bringing my mind to order, reminding myself what it is I wish to experience, to think and feel: grace and calm, a simple appreciation, the pursuit of joy.

Last night I sat on the sandspit looking at a sunset that had no equal, a rendering of intense red and gold by the hands of the master. When the sun set the sky showed a place of pure green light that beckoned like a gateway to heaven. The Polynesians considered such places real. The sunset sky, the glowing towering clouds, spoke of a heavenly land, a source, and a destination. It is hard not to accept this Polynesian geography when looking at the great pillars of violet clouds flanking a plain of pure clear green light, more lovely than any earthly pasture. I watch the sunset sky and drive all thoughts of discontent out of my mind.

In the foreground, Tovi is a black silhouette carrying a spear in the shimmering shallow waters. The sea is perfectly calm, mirroring the dark mass of Yaqaga. The small boys prance on the sand, call out, "*Bula*, Joana, *bula*." Then they giggle. I answer in English, "Good evening," and they answer back, mimicking my tones, "Good evening," and giggle again, beside themselves with delight.

The children are entranced with me, staring with great grave eyes, clustering around me where I sit, shyly touching my arm and stroking the yellow hairs on it. Some are especially devoted and will come dancing out of nowhere to walk by my side, unspeaking, patting my shoulder, holding my hand, as we walk through the village on the way to nowhere.

Village children are strong and able. I watch the small ones, two and three years old, seriously carry small jugs of water from the tank, already proud of their abilities. Seine's daughter, Selai, and small Malé carry a jug of water slung on a pole between them from the water tank. The jug slides down the pole, shifting the weight from one to the other. Selai puts down the pole and turns to Malé, accusing him with her eyes of gross incompetence. Already their actions mimic those of the adults; they are a little couple, relating as man and wife, in the serious business of sustaining life.

Little Malé, not yet three, carries a gallon jug to my *bure* for me to use. When I try to take it from him to relieve his heavy load, he tugs it back and continues. He will spend hours playing in the shallows in front of the house, naked and glossy, looking like a seal, and often comes inside to cast around for treats: an orange, a banana, a cherished cookie. He is growing fat under my patronage. Sometimes when he thinks no one is watching, he sneaks in and makes a swipe at the butter or grabs a handful of peanuts. Big Malé is delighted at his crooking the food and sets traps for him.

When we are eating breakfast, little Malé and Suliana, his three-year-old cousin, come in and inspect our food, watching us eat with great intensity. I give them a cracker with peanut butter, or a cup of Ovaltine, and they leave, unspeaking.

Malé calls them back. "Say thank you, Aunty—*vinaka*."

Their words come out in shorthand. "Anku ti," says little Malé, on the run with his peanut butter cracker. Suliana's voice is more assured: "Tanku ti," also on the run.

Most of the children, and the adults for that matter, live on sugared tea and cassava, fish and breadfruit. They are serious about food; it is their first lesson. Food is life. When invited to come and eat when passing a *bure*, you never say no; that would be insulting the food, saying it's no good. You say *vinaka* (thank you); or, in the dialect of this island, *sa re* (thank you, it's good), and continue. *Mai'ana* (come eat); *sa re* (thank you).

One morning during breakfast I heard children's voices outside our *bure*. I looked out the window and saw Maria and Dimate, Uncle Ame's children, playing "store" with some other children of the *koro*. They had arranged three empty biscuit packages, a tea box, and two empty rusting tins that once held fish on a makeshift shelf in the sand. They were going to the "store" to buy tea and biscuits and fish to have for their breakfast. Dimate, the oldest, sits importantly on the sand, with a folded leaf in one hand and a stick in the other. She is smoothing sand on the leaf, "buttering a biscuit," and hands the sand-covered leaf to Maria. "Come have some biscuits," she says. Maria wants some more "butter," and Dimate adds some sand.

Maria and Dimate almost never eat anything but cassava. I doubt if they have ever tasted real butter. Their thin arms and overround bellies speak of malnutrition. Often at night I hear them singing hymns in high childish voices. It seems to be one of their few pleasures.

I turn to Malé, a catch in my throat, and ask him to give the children some real biscuits with butter. He butters some biscuits, calls outside, and a group of children immediately appears, standing shyly to one side of the doorway. Malé hands out the biscuits, and they run off to share their treats with one another.

The hardest lesson I have had to learn here is sharing. I have been educated by my culture to consider what is mine, mine. Everytime I am expected to share what is mine freely with the members of the family, I balk and get irritated.

If I buy Malé a shirt, I see it on a cousin; my clothes go into the family wash and reemerge on one of Malé's sisters. It's not so much the sharing, but the absolute lack of control over our possessions that confounds me. I have made my own way in life for so long that it is hard to adjust to the communal demands of a large family. Yet it is the extended family, as a whole, that sustains life; and it is the family, as a whole, that must be cared for.

I think that's the reason Fijians never did develop much of a technology for food preservation. Aside from the surplus fish smoked for the Sunday supper, there is no salting or drying of food for later. It would be rude and inconsiderate to hoard food when there are many cousins and aunts and uncles next door who are hungry. So surplus fish or turtle is shared, first with the immediate family, and then in ever-widening circles, until it is gone; and what's the good of hoarding food if one can always die tomorrow. Only the God knows the life.

Cigarettes, kerosene, radios, clothes, food are all shared. Malé goes out to buy a package of cigarettes from his cousin Tovi at the village store. It is a ten-pack of B&H. By the time he returns to the *bure*, there are six rolls left, for he has met four cousins on the way and given each of them a cigarette. In an hour the pack is gone, and we are on the hunt for *tovako*. *Kavuru* (share). While drinking *yaqona*, someone lights a cigarette; five pairs of pairs of eyes watch him smoke, waiting for the decent interval to call *vuru* (share); then the roll is passed from hand to hand, each one taking a few drags before the next *vuru* is called.

Some of the worst times I have had here have been when I tried to buy something and keep it for myself: a gallon of white gas for the lamp in case of emergency, some batteries for the radio in case of a hurricane, some food for later in case of hunger. Here there is no later, only now; and it is considered ungenerous to keep anything for later when it is needed now. This sharing of everything touches me where I am most vulnerable: in the seat of my security. Sometimes I panic when the last egg is eaten. What will I eat tomrorrow? Yet there is always something to eat tomorrow: breadfruit and green-leaf *bele* cooked in coconut milk; *roti* made from white flour, sugar, and coconut. What matters if there is no "balanced diet," no assorted greens and yellows, no serving of proteins and grains? I have my precious one-a-day vitamin pack, enhanced by the glowing athletic European family smiling on the box. It is enough. What is important is that my mind is not impoverished by anxiety about the greens and yellows. Children with boils, with runny noses, grow up, turn into strong healthy adults who carry firewood, chop weeds in the garden, hunt the reef with spears, become expert in the use of the *sele levu*.

The children are not overly protected against harm. Small Suli carries the sharp *sele* to her father with pride, takes a few swipes with it at a tree as she passes, pokes at the hot sticks in the fire. Three-year-olds are sent to the fire for *guto* (burning sticks with which to light cigarettes) by the men sitting in the *yaqona* circle. Selai straddles the board used for grating coconuts, struggles to master the art. By the time she is five or six she will be adept. My efforts to grate coconuts result in bleeding knuckles and ungrated coconuts. The women laugh good naturedly. Nothing is easy if you haven't learned how.

Young Jone spends hours in the shallow water of the reef throwing the spear (*moto*) at imaginary fish. When he can't use the real spear, he will practice with a stick. He will turn into a good fisherman for he will spend thousands of hours learning to throw the spear accurately. He doesn't do this because someone tells him he should, or because someone takes him out to teach him. He does this because he wants to become like his uncles and grandfather: expert with the spear.

Jone is not the least bit interested in attending school. The land and sea are his schoolroom, his grandfather and uncles his teachers. There is a government-supported primary school in Lekutu that the children of the *koro* may attend. Although education is free, except for school uniforms and books, it is not compulsory. Most of the village children go to the school in Lekutu. They must board there, for it is too far to go daily. So they leave after church on Sunday and do not return until Friday afternoon. Family members rotate the responsibility of going with the children, cooking for them, and bringing the food. There is no cafeteria; all of their food must be provided by their families in the village. School might be fun, but there is no reinforcement of learning once they get back to the *koro*; parents are not interested in the exotics of reading and writing; there is too much to do just to stay alive.

The *koro* is the real school, teaching the techniques, manners, and relationships of the Fijian way. Most often, sending the children out of the *koro* to school deprives them of the skills necessary to survive and does not replace them with enough to enable them to make their way in the money economy.

One Sunday afternoon I watched the village children take off in the boat to go to school. Uncle Saimone's son, Kavaia, himself only thirteen, was the boatman. Kavaia refuses to go to school, preferring instead to fish with his father. He is an accomplished fisherman and an expert boat handler. The kids were all sitting in Uncle Saimone's punt, waiting to take off for school that afternoon, and the young daughter of Asenaca, the chief's daughter, was protesting. Screaming and crying, she didn't want to leave the village to go to school. The boat departed, the women watching with amusement from the shoreline, and we could hear Asenaca's daughter, Tima, screaming from afar. Shortly thereafter the boat turned around and came back. Tima had thrown everyone's belongings into the water: clothes, bags, food, all went floating. They had to turn back; and everyone seemed happy. "I guess she just doesn't want to go to school," said Asenaca, laughing.

Slowly we are moving ahead with our housing plans: rebuilding the family's big *bure*—Botoi—and building a bigger house for Malé and me on Vedrala and a new house for us here on Galoa for when we come on weekends for church, for special occasions, or just when Malé wants to drink *yaqona* and play music. Since we still spend most of our time on Galoa—Malé is stuck to the social life of the village and his family—Momo has interrupted his labors rebuilding the big family *bure* here to make us a new *bure*—our own—under the *baka* tree.

The news is that Seine will be back from Suva any day now, and so today the whole family has gathered to complete our *bure*. It is almost finished, the woven coconut *bola* still green. It is a fairly big house, built on the edge of the sea, no more than three yards from the highest tide mark. Shaded by the huge *baka*, it looks cool and inviting. Momo and I have a serious consultation about location of doors and windows. I have not yet learned to defer to his superior knowledge and think I want more windows. He agrees. He will let me learn later, when the cold winds blow from the direction I want my big window to face. Making windows is easy. You just cut the appropriate-size hole in the *bola*, weave the edges together, and you have a window. Closing windows involves finding a suitable-size piece of old corrugated roofing lying around and propping it against the opening with a stick stuck into the sand.

Nei, Ima, Una, big Suliana, Malé, and young Jone are sitting under the *baka* weaving the *bola*, surrounded by a shimmering sea of green fronds. Everyone flocks together, and I take some pictures. Malé leaves his *bola* weaving, ambles into Seine's *bure* and returns wearing my Hawaiian hat, sunglasses, and carrying the spear gun. He poses with the spear gun on the gunnel of the punt, pointing the gun at the still-blue sky. Taking pictures is a highlight; everything turns into fun.

Running to get into the picture, small Malé trips over the *bola* and hits his head on the sharp edge of a coconut rib. The sore over his left eye, which hasn't healed for months, opens and flows fresh blood. He screams and rubs his eye with his sandy hands. No one is alarmed. I wash out the sand and look at the ugly puffy wound over his eye. It is in a place that's impossible to cover, and each day the flies work deeper into the sore. I wish someone would invent a simple antibiotic salve that would repel flies and stay in place. By the time I return to the *bure* and sit at the typewriter, little Malé has forgotten his sore eye and is laughing and running around with Suli.

Malé and young Jone have gone off in the punt to catch some

fish for lunch. I will cook some *dhal* and rice to supplement the hoped-for fish. I am feeding the family today, for they are working on my house, and it's time to start cooking the *dhal*.

VONU

In the old days men returning from a successful turtle hunt were pelted with rocks and green breadfruit and papaya as their boat came to shore. The women of the village gathered on the sand and sang and danced a traditional *meke* as the boat came in loaded with its precious cargo of live *vonu* (turtle). The men in the boat dodged and laughed. If they didn't dodge, they got hit by a rock or a rock-hard breadfruit. If they didn't laugh, well, then they weren't very good Fijian fishermen.

Momo has been asked by the prime minister's son-in-law in Suva to supply him with some *vonu* for an anniversary feast. A request from a chief, and such a high chief, is a royal command. To refuse it would be unthinkable; to perform it is a great honor. So the past week has been devoted to getting ready to hunt the *vonu*, and hunting the *vonu*.

Getting ready requires the drinking of *yaqona*. Everything important requires the drinking of *yaqona*. We took the boat into Lekutu while I grumbled about why I had to spend my money for the *yaqona* and tobacco and benzine when the prime minister's son-in-law was a rich man, and we bought the necessary supplies.

On the way back Malé explained to me that when Momo took the *vonu* to Suva, he would be rewarded in the old Fijian way: with things that are of Fijian value—mats and *tabua* and *tapa*—not money. Okay, I conceded, thinking I might come out of this adventure with a fine mat or a piece of princely *tapa*. I settled in to appreciate this new diversion.

After a full night of *yaqona* drinking, shared by most of the men in the *koro*, and a lot of talking about *vonu* and the ways of getting *vonu*, and the consumption of a carton of cigarettes, Malé's brother Vili and his cousin Kava left in the morning hours. In the dense dark before dawn, they loaded the turtle net into the punt, added a teapot full of hot tea, grabbed a handful of cigarettes; and amid a cloud of good wishes from those still in the *yaqona* circle, set out to hunt the *vonu*.

I went to sleep, thankful that Malé had been spared the expedition by his father, who was, as usual, trying to keep me happy. When I woke and stumbled out of the *bure* into the sunshine, there were two immense turtles lying on their backs beneath the *baka* tree, and a smiling and elated momo saying, "*Yadra*, Joana." I returned an unfeeling *yadra* and sat on the punt brushing away the flies and disliking men, all men, in general.

The *vonu* looked so pitiful in the sand, their great bodies overturned and helpless, soft dark eyes blinking away the unaccustomed flies, powerful flippers carving a hopeless pattern in the sand. I sat there, digesting my bile, looking at the helpless *vonu*, a hundred years of bright water and sparkling sunlight to end there in the gritty sand, almost suffocating under their own weight. As I sat there, young Suli, herself a dainty frightened bit of soft smiles, tentatively kicked at the *vonu*'s neck, and he arched his flippers and sighed. My tears were ready now and the hate rising. Then I saw, way out on the reef, young Jone, making his way to the shallow pond Malé had dug for a baby *vonu*, which he had given me as a pet, with the baby *vonu* in hand, putting it in the pond as a relief from the outgoing tide and the morning's heat. The world was too complex for tears.

"*La'o mai*, Joana (come)," called Momo, motioning me to come and sit beside him under the *baka* tree. "Ima, bring some tea," he shouted, and we sat together greeting the slow parade of villagers who came to view the *vonu*.

Vonu is the sacred food of chiefs, the *mana* of the sea, just as the *vuaka* (pig) is the *mana* food of the *vanua* (land). No feast is complete without the *vonu* and the *vuaka*. No matter how much other food there might be—cow meat and chicken, *dalo*, cassava and breadfruit—there must be *vonu* and *vuaka* to feed the chiefs. Momo was glowing with pride as he explained this to me, with the help of Ima. It was as if he himself had taken on the *mana* of the *vonu* in being able to present these mystic animals to a great chief.

To Fijians *mana* is a reality, a supernatural power that resides in certain persons and objects. It is a gift of God. Chiefs are given *mana* when they are named as chiefs. The higher the chief, the greater his *mana*. Some common people have been given *mana* as a gift from the supernatural world, and with it are able to perform extraordinary feats, just as the famed firewalkers of the island of Beqa were given the *mana* to walk, unharmed, over the burning stones of the *lovo* (the Fijian earth oven).

Sometimes when I look at Momo, I feel his *mana*; his personal presence transcends ordinary men. He is the only human being I have ever met who radiates this quality, a quality I have read about in literature of Polynesia, but never before felt. Sitting in the shade that morning I felt that presence in him as surely as I felt its absence in everyone else.

Momo's father, the first Jone Varawa, was in his lifetime one of the few whose special responsibility it was to hunt the *vonu* at the behest of the chiefs of Lekutu and Bua. Then the hunting of *vonu* was a profound ritual. A special *yaqona* ceremony preceded the hunt, and strict silence was observed in the boats. When the *vonu* had been captured far out on the fishing grounds, the great conch shell (*davui*) was sounded by the men in the boat, sounded continuously as the boat made its way back to the island. No talking, no laughing, only the hollow eerie sound calling across the waters to announce to the villagers that a *vonu* had been captured for the chief. All through the long journey, while the fishermen paddled in silence and the *vonu* struggled in the alien air, the great conch blew. Somewhere still, in this village, hanging from the rafters of an abandoned *bure*, is the old, old net that was once used to hunt the *vonu*.

We spent the day drinking *yaqona* again, the men in the shade of the *baka* softly strumming on the guitars, as the *vonu* signed, slowly turning in an impossible universe. The music was sweet and the men kind and thoughtful. Malé was happy and soft eyed, lighting my cigarettes and pinching my thigh. The *yaqona* tasted good—sweet and calming—and I drank *bilo* after *bilo*, listening to the slow talk and gentle music, propping up my resolve to stay here and dreamily thinking of getting married, of the *tapa* I would wear at our wedding, of the *salusalu*, the garlands of flowers we would wear around our necks, of the *vonu* we would feast upon after the wedding.

That night Malé wanted to go out with his cousins for more *vonu*, and I agreed. When he left, he said, "Goodnight, Joana," and kissed me on the lips. When I woke during the night and turned, he was not there, and I reached to the emptiness, "*Isa*, Malé, *isa*."

They left around eleven o'clock in order to paddle to the fishing grounds to save the precious benzine. They paddled until five in the morning. As the sun came up, they were waiting in the shallows near the deep sea. They saw two big *vonu* swimming in the shallows, and Vili and Kava quietly got out of the boat and spread the net between them. Then Malé in one boat and Toa and Mone in another stationed themselves some distance behind the slowly swimming *vonu* and began beating the water to drive them into the net. One *vonu* swam directly toward the net and in seconds was entangled. The other turned and made its way toward a gap between the net and Toa. Malé gunned the engine of the punt and gave chase to the escaping *vonu*. With all of his considerable strength, he threw the heavy turtle spear at the back of the *vonu* and dived into the water. He couldn't see anything because of all the blood, the sea a bubbled cloud of red; and in the froth he grabbed the flipper of the immense beast—shark, no shark, no matter—and, holding onto the flipper, swam as best he could, with one arm fighting the *vonu*, toward the drifting punt. Toa and Mone came up in the other boat, and the three of them managed to haul the huge turtle into the boat. Another fifty years of bright water sighing. Malé was clear eyed and ecstatic when he told me the story in the *bure* as I sipped my coffee and listened to his tale.

After breakfast the men butchered one of the *vonu*, because it would die before they could take it to Suva. Before they carved into the *vonu*, they all stood in the shade of the *baka* and propped up the four huge *vonu* for a picture. I took the picture, looking through the lens at the man I wanted to marry.

QUESTIONS

We have just boiled some water on the Primus and made Ovaltine. It is the middle of the night, and we are sitting companionably by the light of a kerosene lamp at a low table next to one another. Malé is copying Fijian songs into his songbook with grave attention. I am writing and listening to the soft rain sounds on the *bola* of our newly finished *bure*. We are, finally, in our own home together.

The tape is playing a sad Fijian song, a song about love and loss. "*Sa oti*, *sa oti* (it's finished, finished). The song discomforts me somewhat because last week Malé and I had a crisis over a former girlfriend who was visiting in the village. I found a letter she wrote to him about an evening they had spent together while

I thought he was drinking *yaqona*; I remember that night for having awakened in the middle of it and having gone in search of Malé, who I assumed was still drinking grog somewhere. I prowled through the sleeping village hunting for a light or the soft murmur of voices that would tell me someone was awake, but I couldn't find Malé anywhere. My fear rose like a flood tide, and I woke Momo, who went through the village calling Malé. A few minutes later Malé stumbled out of the darkness with a barely believable story about falling asleep in a *bure* at the other side of the *koro*. Then I found the letter.

There were tears, and threats from both of us, until we reached some sort of equilibrium and the promise from Malé that it was *sa oti* (finished). Now, listening to the sad words of this song, I read into his interest something ominous, and then remember that my meditation of the last two days has been on the subject of self-inflicted pain.

I am trying to understand how the mind heals itself. I believe that my body can heal itself if given the right conditions of care and cleanliness and time. I see what happens to my body when I scratch an inoffensive mosquito bite. By irritating what is already irritated, I inflame it. It then gets swollen and painful and forms a boil in the heat. The boil takes a long, long time to come to a head, a time of pain and discomfort. When it finally does come to a head, it discharges a disgusting load of blood and pus for days, slowly draining and crusting over. If I leave it alone and keep it covered so the flies can't eat on it, it will slowly heal, a process taking from three to six weeks. If I start scratching when it is healing, it will break out in an adjacent area, and I will have a new boil. If I had left the bite alone in the first place and kept my skin clean, there might never have been a boil.

Is my mind like that? Do I scratch and scratch at trouble,

inflame the source? How do I control my memory so that it doesn't continually irritate me? Can my mind heal itself?

Two days ago I went to Labasa by myself to shop and be alone for a few hours. I spent a night in the luxury of a hotel room, reading *Zen and the Art of Motorcycle Maintenance*, trying to understand my relationship to Malé. Cut off from my own language and culture, my reactions to Malé's actions are very often touchy and emotional. Nothing around me seems to support my better sense of myself; my competence is all in another area. I have traveled around the world by myself, and on Galoa I cannot find my way around in the bush; I get lost going to the garden. My independence needs telephones and cars and the materials of another culture. Here I am weak, unable, and, very often, unstable.

Reading that book reminded me of the usefulness of reason. My own language reassures me, for here I am constantly surrounded by foreign conversation that makes no sense and keeps me from having even a sensible interior monologue. Here in Fiji, much more even than at home, I need to think. And here, too often, I cannot, and do not, think.

Very often I am unhappy, and I blame my unhappiness on the conduct of others. My mind is constantly searching for flaws, for bites to scratch. If I have a lover, it is all the easier, because I can blame him for not loving me, or for not loving me enough, or for any number of possible bad things that cause my unhappiness.

In Labasa I thought about my relationship with Malé and confronted the fact that since I am often unhappy and he is most often happy, and since his happiness does not derive from me, but from his own sources, I am jealous. The only way I can get his attention is to blame him for my unhappiness and to point out his faults. This, of course, makes him unhappy, so I have succeeded in making both of us miserable. Then he rejects me, goes elsewhere to be happy, so I have proved that I was right about him—he doesn't love or understand me.

This is a classic no-win situation: I lose by winning. I see it more clearly with Malé than with any other lover, for he has little skill or interest in the complicated psychological game Americans often mistake for relationship; and instead of participating, he just gets dense and goes outside. Joking with cousins is infinitely preferable to trying to answer the elusive question—do you love me? I guess it is exactly Malé's inability to cooperate in this mental morass that makes me see it more clearly than I ever have before. It is a valuable lesson, perhaps one of the most important in my life at this time.

Maybe I instinctively understood this when I chose to come back to Fiji and try to build a life here in what sometimes seems like an incomprehensible—or worse—situation. Maybe I understood that it is necessary to take myself out of the known and perfected complications of my own culture and see myself in the stark mirror of an entirely different one.

I believe that if Malé and I can construct a love together—for it is definitely a construction and not a gift—then, in fact, it will be love, and not the blind panic I have previously experienced under that name.

So here we are, at home in companionable quiet, listening to the rain continue its ministrations. Sonny, our dog, is a warm golden presence beside us, and Malé's beauty a dark shadow on the intricate brown-and-black-and-white pattern of the *tapa*.

Our *tabua*—heavy, polished, inscrutable—contains its own memories. It hangs in the central place of pride and honor against the *tapa*, reminding me of my promise to marry. The crisis has passed, the tape has finished its sad song, and Malé has gone to the coolness of our bed. I turn down the lamp and follow him. It is time to sleep. "*Moce* (goodnight), Malé."

⁂

It rained for the rest of the night, and in the early morning, when I went out under the umbrella to pee, the sky and sea were the same gray mist, scarcely distinguishable. This is a rain that will last for days and days, a slow steady gray enveloping rain that says rest, be peaceful, there is nothing to do, there is no need to tear around in search, stay still and be warm and listen—for that is what is offered today.

Malé lets me sleep a long while after I come in from my early-morning peeing: a delicious, comfortable sleep. What was bad has passed between us, leaving a gentle warmth, a forgiving and closeness that did not exist before. The wheel has balanced and rolls smoothly.

Malé calls me, "Wake up, sweetie, here's your coffee," and my eyes open to his smiling happy face. The lamp is open and glowing. While I drink my coffeee, Malé plays his guitar and sings. His voice is warm and rich, very beautiful.

I take a book and lie in the dim light of the *bure* door. Through the coarse woven *bola* of the makeshift shutter, I see the gray sea and sky. The bright colors of the boat, floating in the mist, remind me of the dream time. I read for a while, but the complications of thought bore me. On the ground just in front of me, growing almost inside the *bure*, is one tiny green plant brightening the rain-saturated sand. A drop falls on the plant, it trembles.

Inside the *bure* it is dark—from the gray sky, the closed doors. The typing table is a clutter of papers, dead ash of mosquito punk, cigarette butts, my notebooks and books, a silent radio, and the bright glow of the lantern. A spray of spider ginger buds that I picked a few days ago has blossomed; dead petals form a brown corolla around the bottom of the newly opened pink-white flowers. Their perfume is heavy and intoxicating. Malé's guitar leans against the *tapa*; a pile of newspaper-wrapped *yaqona* bundles waits on the new mat.

Our new *bure* is reassuringly beautiful. It is green and fresh, the logs fragrant, their bark unmarred by the intrusion of termites and spiders. I am living inside a big fresh green basket, and it enchants me. The architecture of the *bure* is calming: the lean of the poles, the textures of the woven *bola* and mats, the human motion contained yet not frozen into the requirements of the straight edge. Here I am surrounded by the evident work of human hands in concert with the organic materials of this earth—coconut palm leaf, pandanus leaf, tree trunks and saplings, vines. We are at home under the *baka* tree; from the opening of the doorway I see its huge trunk, the hanging roots, the drifting leafy branches. I don't know where the *tēvoro* goes in the rain. I suspect he, too, is gentled and resting.

Malé has gone into the bush with Toa to pick the breadfruit we will eat this afternoon. Before he left we talked about what we will have for dinner. We have not yet blessed this *bure* with a family feast, and today we shall. Malé will buy a chicken from a neighbor, kill, and clean it; and I will cook it on the Primus stove with potatoes and onions brought from Labasa. That is what today is for: to rest, to cook and eat with the family, and to be thankful for the life and for the love.

DEATH

Rain all day yesterday. Drenching gray rain from morning to night. We went to Vedrala to cut grass to thatch Momo's big *bure* in Galoa. On our return, Momo was standing at the shoreline. He silently threw a stone in the water in front of the boat. It was the signal of death.

Now I hear the wailing, the deep ceremonial sobbing that announces death. Two cousins have come from Tavea with the news that Malé's young nephew died yesterday—the day we were cutting the grass. The boy was at school, sick for two days; he couldn't eat. The children told the teacher he was sick. She gave the children some tablets for the sick boy. Finally the teachers took him to the health center. The doctor said there was nothing he could do. The boy died in the van, in the driving gray rain, on the way to the hospital in Labasa. The family draws together.

The women wail, sob; the men sit silently, tears in their eyes, and listen to the sobbing. The sobbing is infectious; it draws my tears. This morning I awoke with all manner of minor complaints—a refusal to love Malé, a questioning again of motives. My complaints vanish in the universal sound of grief. I look at the pattern of the mat, play with a few grains of sand, see clearly the signs of material poverty: the worn clothes, the crowded and disarrayed eating house that is also a sleeping house for seven people, the unfinished mats piled in the corner, the spare food.

This morning I asked Malé questioningly, "What no *bula*?" No morning greeting of life and health?"

He looked at me blankly, "No *bula* no life." No *bula* now for the small boy.

Momo organizes the family. Malé and I will take the boat to Nakadrudru and buy the sugar, tea, rice, and flour to take to the family in Tavea. The family here will gather the *tapa*, the mats, the *tabua*; and we will all go to Tavea to sit with the family, to sleep there tonight and drink *yaqona*, and tomorrow to bury the boy. Our plan for today was to have been to return to Vedrala and cut more thatch. It is hard work. Burying the boy is harder, and I never even met him. He was a twin.

⧋ ⧋ ⧋

Mosi, mosi. It hurts. This really hurts. We have come to Tavea; Momo and Nei will come later. The day is unsettled—heavy skies, not yet raining, but promising big rains. Malé, Vili, and I are sitting in the tiny house of Malé's uncle Malé, the man he is named after. I look at four woven bamboo walls, some mats on the floor, one dark cramped bed in an airless curtained corner. Not much else except the sacks of sugar, rice, and flour that have been brought to the family as funereal gifts.

Uncle Malé is quiet, holding his grief; tears come to his eyes then subside, and he makes an effort to talk as if this were still the ordinary world. The talk is of the manner of the boy's death. Jack, eight years old, lay in the dormitory room of the school for two days, vomiting, soiled by his own diarrhea. His twin brother sat up with him all night while the other children slept and massaged his forehead, his aching suffering head, already tortured by his coming death. The family talk is that the teacher was to blame. I have no idea if anyone is to blame, but I see this frail man, Momo's older brother, and my heart goes out to him. From a house nearby the wailing of the women rises and subsides, alternating with the calling and laughter of the young people. I say nothing, light Uncle Malé's cigarette, hold his hand, drink my sugared tea, and look out the doorway at the bright pink and yellow fluttering of two small girls. A tiny child appears in the doorway, looks at me gravely, then begins to scream. I am thinking of the twin, of the boy who just lost half of him, and what he must be feeling.

I ask Malé about the boy; he goes outside and calls him into the house—a thin, shy boy with enormous appealing wounded eyes, and arms and chest covered with a terrible scaling skin disease. Malé and I talk about taking the boy to the health center for some medicine for his skin. The boy doesn't care. I take an orange from my purse and roll it across the mat to him. He looks down at it and doesn't touch it. Vili tells the boy to go get a knife. He leaves and comes back with a knife without a word. Vili peels the orange expertly in one long fragrant spiral. He hands the peeled orange to the boy, and the boy sits, holding the peeled orange in his hand. I feel hopeless, have no idea where to begin. I smile and look at the boy; he looks back, unsmiling.

Outside, women are coming and going, doing the cooking for the funeral. Inside the house it is quiet; we drink our tea and talk softly. One of Malé's aunts comes in and kisses me in the Fijian manner, sniffing the side of my face. She smiles and pats my hand, saying *vinaka*, *vinaka* (thank you) over and over, then talks to me in Fijian. I have no idea what she is saying, but smile and rub her thin shoulder. Malé asks me for my purse, says he wants to go buy some cigarettes, and I give him ten dollars. He goes out, telling me to wait there, he will be back. We sit together—Uncle Malé, his sister Sereima, and I—and smile at each other, and I speak childish phrases about how hot it is. It is stifling hot: no wind, the heavy tense heat of bad weather to come.

After a short while Malé comes back and divides the carton of cigarettes into two five-pack bundles. One bundle he gives to his aunt to take to the *vale ni mate* (house of death), and the other he keeps to take to the women who are working in the kitchen. He tells me to take my camera and go with Aunt Sereima to the *vale ni mate*. I follow her on the muddy path past the crowded houses of the village, thinking that this must be where they are sitting in wake over the coffin. We walk up to a large wooden house and look inside.

I am stunned by the beauty before me. The floor is covered with exquisite bright wool-fringed mats, ceremonial gifts, laid in display, a rainbow of jewel colors surrounding the patient weave of the *voivoi*. At the far end of the room, laid on top of the mats is a soft brown *tapa*, a *tapa* made on Tavea. Sitting cross-legged in a rectangle around the mats, leaning against the wall, are sixteen women. Three of them are in black; the others are wearing dark *sulu*. I am asked to come in, and, circling the room on my knees, greet each woman in turn, sniffing their cheeks, saying *bula*, *bula*, *vinaka* (good health, good life), touching a shoulder, caressing a hand as I go. One very old woman greets me with tears and holds onto me. When I have gone around the room, I am asked to sit in the place at the far end, next to Aunt Sereima. There are no flowers, no coffin, only the bright bloom of the mats and the soft earth reminder of the *tapa*. Malé's aunt takes out the cigarettes and hands them to the woman on my right. She is dressed in black. She receives the cigarettes in a manner similar to the *yaqona* ceremony, clapping her hands three times (the *cobo*) and making a short quiet speech of thanks. Then she carefully divides the packs and hands each woman her share of the cigarettes. Matches are passed around, and we all smoke. While we are smoking, Sereima tells the women the story of my opening my purse and giving Malé the ten dollars. She tells it at great length, imitating how I opened the zipper on my wallet, and tells about the cigarettes that were given to the women in the kitchen. I do not understand many of the words, but get the idea. The women listen with great appreciation and warmth, nodding and saying *vinaka*, *vinaka*. The talk changes, becomes something else that I do not understand, and I look at the floor, at the mats and the *tapa*, and at the women.

Then from one of the women in black a wail comes, a deep moaning wail; and it is taken up by each woman, some with much more intensity than others. *Oilei*, *oilei*, *oilei*, repeated over and over, a crying from the womb for the lost child. I am carried into it, the way a swift current sweeps away the innocent, carried into the dark swirling river of uncomprehended death. While the women wail, small children float in and out of the doorway, wisps of yellow, pink, and orange. It begins to rain, a heavy rain, and the doors are closed. We sit in the darkened room, surrounding the brilliant bordered mats, and the collective grief comes out and is given back to the universe, accuses the heavens, mingles with the thick gray rain. *Oilei*, *oilei*, a dialogue with death. After about five minutes, the wailing begins to subside, and the women dry their tears, sniff back their phlegm. Aunt Sereima hands me a square from an old towel, and I dry my eyes and stare at the mats.

Slowly the talk resumes. Something ordinary is said, then answered. I look at the colors on the floor, fringes of intense pink, yellow, blue, purple, green, and red wood woven into the borders of each mat. I have never seen anything so beautiful in my life.

Una comes to the door and motions me outside. The rain has stopped. I ask Una where Malé is, and she points to him standing under a breadfruit tree nearby. I go to Malé and ask what he wants. He wants me to take my camera and go to the shoreline and take pictures of the Tavea church choir. On my way I meet Ima, and we walk together and sit under a tree on a makeshift bench. The choir is standing in the sand: about thirty young people, dressed in white *sulu* and shirts and blouses. They are singing a hymn, looking out to sea. The tide is low, and before us stretches the shimmering flat plain of silver water. On the shoreline the chickens peck in the sand. A pig walks by. The singing is soft and sad, perfectly rendered. In the distance I see a boat slowly being pulled through the shallow water by a number of people. Now I understand. It is the *waqa* I heard the women talking about. *Waqa*, the canoe, the boat, the coffin—the container.

The dead boy is coming home in his coffin. The scene is achingly clear: the flat gray distant sea, the white *sulu* of the choir, the incredible soft sad singing, the slowly moving boat coming closer and closer to the shore. Six men walk to the shoreline, older men, dressed in shirt and *sulu* They stand silently, hands clasped behind them, waiting for the boat. Then in the distance I see the coffin unloaded. It is so small.

The six men wade into the water; they present a *tabua* in a ceremony to welcome the dead boy back to the island of his birth; then the pallbearers pass, strong young men carrying the little coffin. As they pass, I take a picture of the coffin for the family. It is covered with something brightly patterned. There are no flowers.

The choir and the villagers fall into line behind the coffin, and the funeral procession passes. Ima and I follow them up to the grass in front of the church. The coffin is laid on a bed of mats arranged on the grass, and everyone sits down. I look for Malé and find him seated in a group of young men. I go over to sit with him, but feel out of place, feeling that I should not be sitting with the men. I ask if it's okay, and Malé says to stay.

A man with a deep strong voice prays. Another man prays. The voices, assured and resonant, carry in the quiet air. Another man kneels, holding two *tabua*, and makes a speech. I listen to the words rolling in the heavy air, the ritual ancient words.

Three claps (*cobo*), then three again.

The truth is spoken.

Malé looks up; then he looks at me. The rain is coming. It comes seconds later—an outpouring, a drenching rain. The floodgates of heaven have opened. Everyone runs for shelter. The people sitting close to the coffin rush into a tiny house nearby. The men pick up the coffin and crowd in after them, a disorderly crushing assault with the tiny coffin. It is unbearably intense. From inside the wailing begins. *Oilei*.

Malé and I and several other people seek shelter in a small shack with a dirt floor and a table faced with mismatching scraps of old plywood, and on a shelf behind the table, or counter, are six small cans that once held baking powder; there is nothing else. It is the village store. We sit in the dark airless shack and listen to the wailing. Malé leafs through the account book of the store. A young boy, the storekeeper, comes in and tries to get us outside so he can lock the store. It is absurd. Malé talks him out of it and goes back to looking at the tattered notebook of small transactions. Behind the counter the storekeeper rearranges the six empty cans on the dusty shelf. I sit on a bench made of sheets of tin covered with plastic rice bags, listen to the rain, and look at Malé. After a while the rain lets up a little, and we dash back to Uncle Malé's house.

Malé sits down on the mat, smokes, says nothing. I lie down. I am really tired, feel the weight of life and death settle into my bones, close my eyes, am drifting into sleep when Malé shakes me hard, "Get up, get up, move."

Two men crowd through the doorway carrying the frail limp form of Uncle Malé. He has fainted in the cramped airless room, devoured by his grief, listening to the wailing and looking at the coffin of his dead nephew. I feel that he wants to die. We push away the corrugated iron sheets propped against the doorways, and his niece gets an old shirt soaked in cold water, presses it against his thin chest. Uncle Malé's breathing is far away, coming from the grave. His wife comes in and holds his head and fans him. I ask Una to fan him, and I take hold of his cold feet. I want to remind him that he still has a body, for I feel he is lost in his mind and will not come back.

I remember reading that the old-time Hawaiians would restore the dead by forcing the soul back through the big toe, and begin to massage his big toe and call, speaking under my breath, come back, come back. Malé sits in stony silence, expressionless, looking at his uncle lying in front of him. After about ten minutes of fanning and massaging and stroking, Uncle Malé shows signs of returning. He opens his eyes, unseeing, then slowly focuses. He is here with us in this room. He is not in the grave. He sits up, and someone brings tea. His wife holds the cup to his lips, and he sips, looks around more intently, and sighs. Soon he is sitting up and drinking his tea. He says something to Malé that I don't understand; then he looks at me. His eyes flood with tears, and I hold his shoulder, feeling the bones under his thin shirt. I want to say cry, cry, cry, and understand he cannot. His tears are held in, strangled, controlled. His wife wails *oilei*; and Uncle Malé sits, sipping his tea, staring.

I massage his head, stroke his temple; and a low moan floats into the room. He is moaning like a child who is being comforted, and I stroke his eyes, his hair, and long to take him into my arms and rock him like a baby. Finally he looks at me and says, "Thank you, Joana."

We sit and sip our tea. It is getting late, and I want to go home to Galoa, for I am afraid of going back in the dark in the driving rain. I ask Malé what he wants to do. He answers by saying nothing, which is his way. Someone brings news that there is a tropical cyclone coming. The name of the cyclone is Martin. It is my dead father's name.

I worry about the weather, want to be home if a hurricane comes, ask Malé again. He looks at the mat. I know he wants to stay on Tavea and drink *yaqona* with his cousins and uncles. I ask again and explain that if bad weather comes, Momo and Nei will be home without the boys, for Vili and Qare are with us. After a time of quiet convincing, Malé turns to his uncle and says we will go home tonight and come back tomorrow for the burial.

The sky is dark gray when we finally get into the boat—Malé, Vili, Qare, and I—and set off for Galoa. Una and Ima stay behind to help the women with the cooking. The four of us sit in the small frail boat under a heavy frightening sky and say nothing. The hurricane is coming.

BOTOI

Sheaves of fragrant grasses cut in the rain on Vedrala lie waiting around momo's big *bure* on Galoa, great bundles of green—shining, heavy, intense. He is rebuilding his family house, a traditional house built on the old family foundation, a house of cane and vine and grass, complex and beautiful. Momo rebuilds Botoi in the old style, in the way of his forefathers, just as the house retains its name, the meaning forgotten, bestowed by someone long ago at the moment of its first completion. The foundation (*yavu*), a raised platform of sand, bordered with stones, was given to Momo by his father with the same ceremony that his father received it from his father. *Dela ni yavu* (on top of the yavu) is the place one comes from, the homesite. The line remains unbroken; from father to son the ritual, the technology, continues.

Botoi is the biggest, most traditional *bure* on the island; if you did not know, you would assume it is the house of the chief. The chief of Galoa lives in a big old wood-frame house, but could, if he wanted to, command the village to build him a traditional *bure*. The Fijian *bure* is slowly fading out of the culture. It is easier to build a wood-frame rectangle with an iron roof and install ready-made louver windows and wooden doors than to gather and shape the materials of the bush into the traditional architecture. Easier, but it takes money. A big house like Botoi can be built for the cost of: paying someone to cut the bamboo on the mainland, the nails, the benzine used to make the trips necessary to gather the materials, and the *yaqona*. All in all, about two hundred dollars. A big wooden house would cost about five thousand dollars. But the sense, the feeling, of the two houses is as different as the cultures they represent. One is impersonal; the boards, the nails, the corrugated roofing have no meaning—they are simply units. The traditional *bure* is replete with meaning and memory.

Each post, each sapling, the bundles of grass, the lashing gathered from the wet mountain valley above Nakadrudru, carry the scents, the associations, the quality of their gathering. Here are the rainy days on Vedrala cutting the thatch, the trip to the mainland to bring the huge piles of bamboo, the slow funny ride back towing the big bundles behind the boat, the days spent shaping the posts under the *baka* tree, and the sweaty afternoons spent pounding the bamboo and weaving it into wall-size mats. This post was that tree; we remember the ferns, the shade, the scent of ripe mangoes the day we went into the bush for this post. All are woven together to create a living structure: a home.

Botoi is a house of surpassing loveliness, with a high soaring roof intricately supported by the curving limbs of once-living trees. It reminds me of a cathedral, buttressed equally by belief and structure. I sit inside, tucked in a corner where I won't be in the way, and watch as the thatch is lashed to the roof framing. Malé and his cousin Ratu run on the interior framing like gleeful cats, surefooted and exuberant, shout acknowledgment to Momo and Vili, who are working on the outside scaffold, laying the thatch. The thick heavy bundles are laid over a framework of woven bamboo strips, pounded, and compressed into place beneath long sticks, and at the moment of maximum compression, tightly lashed from the inside by the joking cousins.

To keep it going and to coordinate their actions, Momo shouts from outside, "*Vaka rau, vaka rau.*"

They begin to chant, momo outside: "*O va'a rau* (get ready), *O va'a rau; va'a e dredre* (tie it), *va'a e dredre.*"

Malé and Ratu inside, "*Yea a a* (yes)."

Momo starts to pound the thatch and chants, "*O awa ra.*"

Malé and Ratu answer, "*Na awaca* (your family's no good)."

Momo chants, "*E tiko na sovu* (the octopus stays in his house)."

Malé and Ratu answer, "*Mai NaBala* (in Naivaka)."

They repeat the chant until the thatch is tied, then shout lustily, "*Yea a a a.*"

It's fun. It sounds wonderful, transforms drudgery into play, reinforces the strong work with the strong voices. The air around Botoi rings. The *rau* (ceiling) takes shape; slowly the sky closes, becomes inside, as sheaf after sheaf of heavy long grasses is pounded under the rhythmic beating of the stick and the answering chant. The prosaic becomes meaningful, has a luminosity.

Botoi smells sweet, is alive and flourishing. I wander outside. Momo and Vili have reached the roof peak, are lashing the *rau* with great hanks of brilliant rust red string beaten from the mangrove root. Ima stands feeding the string, hundreds of yards neatly piled in a circle at her feet. She maintains a careful tension, is relaxed and attentive. This strong young woman, stronger than most men, with the straight body and the open crooked smile, is serious, always serious about her work. Uncle

Tevita, Momo's older brother, digs in his pocket, looks up at the roof, and lights a cigarette. The family members are working together; all are knowledgeable.

I go to our *bure* to get my camera. Vaseva has cleaned our house. The bed is made, the pillows plumped, the table is cleared of cups and bowls. In a glass jar on the table with my books and typewriter is a spray of budding frangipani; its thick perfume welcomes me.

Malé has decided that Vaseva should be our housegirl, should take care of us, which she does with quiet and graceful attention to detail: fresh flowers or greens arranged in a jar, a small collection of leopard cowry shells lined up on the *tapa* tablecloth, the mat swept clean of dirt and sand. For all of this daily care, she wants a watch, a nice one, not the junk kind that goes blank after two weeks, leaving only the black plastic band as decoration. She brings our tea, washes our clothes, picks up after us, with a soft smile, a calm dignity.

Momo and the men have almost finished binding the roof peak; this is the final moment of the house, one that will be celebrated with loud chanting and whooping. I hurry outside with the camera, and they all pose, perched on the ridge like long-legged birds, gesturing and joking. The last length of string is wound over the bundled thatch, tied off neatly, and Momo shouts as loud as he can, drawing out the vowels: "*O O O I. Va'a rau. Seru a belo* (the heron sneezes)."

All answer, equally loud: "*I O, O O O I. Yeah, yeaaah, yeah!*"

Botoi is finished.

Momo is excited and happy. Months of long hard labor have ended. It is time to rest. The *tanoa* is brought out. It is time to drink *yaqona* and tell the stories of houses, to enjoy the admiration of the young. He has earned it. Late into the night the singing and laughter illuminate the *koro*.

In the morning nei and the girls carefully arrange the new mats on a clean sand floor billowy with fresh dry *sāsā*. The wooden bed frames with the built-in headboards are brought in, placed in the far dark side of the new *bure*, covered with intricate mats woven by Nei and Suli that hang over the long edge of the beds and create patterns of brilliant colors in the cool. A huge prized *tapa*, a gift from Nacā, Luisa's husband, is hung on the wall behind the beds, and long narrow *tapa* are tacked onto the great crossbeams that span the house. Under garnish of mats and *tapa*, Botoi blooms, has an intrinsic elegance.

It is just in time. Sunday is the ceremony for the opening of the new church, and tomorrow the chief of Lekutu, a relative of Nei's, will come. This is the house he will stay in during his visit. It is a house suitable for a chief.

NEW CHURCH

The church is opening, and a virgin stands at the locked door holding a white pillow embellished with a red ribbon. On the pillow is a pair of common scissors. Across the locked door is a blue crepe-paper ribbon. The virgin is the daughter of Uqe, the principal builder. She is shy and nervous. She is wearing traditional Fijian dress: a *tapa* bound across her small breasts, a *tapa* wrapped around her waist, another *tapa* gathered into a short overskirt, and a ribbon with a white cowry shell bound tightly around her young throat. She is decoration only. The real power is in the men, many of them ministers and chiefs who have come from nearby villages to participate in this ceremony. The virgin stands silently, holding her white embroidered pillow, while the men make speeches, pray, and present *tabua*. The church choir from Tavea, dressed in white, sings hymns.

Not ten feet from the new church the old wooden church sits gutted and bare. There is a strong wind, and the old church leans away from the wind. It seems ready to collapse; loose boards, old decorations, paintless scallops flap and creak. Someone comes and props two boards against the old church. It is abandoned in favor of the stained-glass windows, the wall clock that chimes the hours, the newly varnished and reconstructed pews, and a prized blue chair, the gift of this village. Flanking the pulpit are two cut-velvet rugs depicting the Last Supper. Below the stained-glass window hangs a small out-of-scale *tapa* and a *tabua*.

I stand watching the ceremony. The minister cuts the blue ribbon and puts the scissors back on the white pillow. The virgin steps back; then everyone files into the church in the proper order: first the ministers, then the chiefs, then the visiting Australians. After them come the white-garbed choir and the rest of the people. I walk away from the church looking for God. Down on the shoreline, floating on the incoming tide, is a single great pulpy purple jellyfish. The jellyfish drifts, dead in the shallows.

After the ceremony and the prayers in the church, everyone comes outside. The men drift into the village *vatu nu loa* to drink *yaqona*; the women return to their work of cooking the food. Last night, down on the beach, the men dug two *lovo*, filled them with wood, and covered the wood with stones. Now the yams and *taro* cook on the hot stones inside. A cow has been butchered, hung from a tree, and hacked into pieces, and is cooking in a number of pots. A white pig lies tightly bound to a bamboo pole next to the *lovo*. A man stumbles over the pig and in rage throws a coconut with all his might at the bound pig. The pig cries. I wander through the village, looking for something to point the way to grace. I find people laughing, eating, and drinking *yaqona*. The fires smoke and sputter.

There are arguments over the distribution of the mats. Many are angry because they feel their contributions have not been appropriately rewarded. Saimone's big white pig with the convivial tail is tied to a coconut tree next to the *lovo*. I remember her when she would cool off in the sea, a great white comfortable presence in the heat. Her two companions, the brown pig and the black pig, have already been slaughtered. The main concern is whether there will be enough food to feed the village and the many visitors.

All day and all night and the next day and the next the men drink *yaqona*. Their eyes are bleary from lack of sleep, their motions uncertain. They spit and look comatose. The women cook the food and wash the dishes and set the table and cook the food again, their constant labor relieved by gossip and laughter. Late in the afternoon the women gather next to the *vatu nu loa* to perform a *meke* (the traditional dance).

They file roguishly out from behind the chief's brother's house dressed in their best *sulu* and *jaba*, red hibiscus and star-like white frangipani blossoms adorn their hair. They have tied big fluffy ruffs of shredded banana leaves around their wrists and necks and look festive and archaic, yelling at one another and joking with the audience. A *meke* has been composed for the occasion, telling the story of the church, the motions contained and graceful, bodies straight and synchronized. Their song and dance is accompanied by the intense rhythms of sticks beaten on short logs, a complex high-pitched clattering of intricate design. As they dance, members of the audience come forward, tuck money into their clothes, cigarettes behind their ears, drape lengths of cloth around the necks of the dancing, singing women. I see the power and beauty of the women, young and old, as they dance and sing. All too soon it is over; the Australian woman and her party, the principal guests, have to leave. A flower opened, adorned the day, and quickly dropped its petals. The gaiety of the women hangs in the air for a few precious moments, then dissipates. The women disperse, pull off their ruffs of fresh green, and leave them lying scattered on the grass. They go back to cooking and cleaning.

Inside the *vatu nu loa* the men laugh and drink and spit, and talk about *tovako*. Malé's main concern is whether or not he has a cigarette. He seems distant and alien, absorbed into the world of men.

I leave the drinking, laughing men, and the working women, the memory of the trussed panting pig, and walk over to the new church in the semidarkness. The door is locked. There is no way to go inside.

Late that night the stars glitter in a cold windy sky. The village is asleep, exhausted and glutted after five days of feasting and *yaqona* drinking. The last decorations have been removed from the temporary eating house; the *lovo* are heaps of dead embers and rubbish; the new church is dark and silent. The ceremony is over.

NOT LONG AGO

Malé and I are drinking coffee and smoking home-rolled cigarettes in our *bure*. We are peacefully together, after days and days of bad feeling and distance. The radio tells a story of riots in Suva in 1959. I ask Malé about the riots; he has never heard of them, confuses the story with stories of people being eaten. I tell him that was long, long ago, more than one hundred years ago; and he tells me that his grandmother was on an island when a boat came with people from another village to kill them.

The shadows flicker in the kerosene light, and Malé's eyes glitter as he warms to the story. One of his grandmothers was already married; it was not so long ago. A message had come from Lekutu warning the people not to go out alone, for the chief of Lekutu was building a great *bure*, and it was feared that someone might be killed to bury under the house pole. His grandmother forgot to tell his grandfather about the warning.

They had gone to Naniqaniqa, an island nearby, to fish and pick shells. At night his grandmother and grandfather were sitting by the fire, telling stories and eating fish. The sea was mirror calm and silent, shining in the bright moonlight. In the distance they saw a big boat coming from Macuata; they could hear the engine from a long way off. As the boat came closer, they knew; and Malé's grandmother remembered the warning. It was the people coming to kill them.

Malé's grandfather ran to the fire and heaped it with brush to keep it burning high and bright—to confuse the people in the boat and make them think they were still on the island. Then they ran to their boat and pushed off silently into the shimmering sea. Malé's grandmother was terrified, sobbing violently. His grandfather told her to be quiet or they would be heard. Soundlessly they paddled inside the reef, where the big boat could not come, and then circled around the island to the safety of the mangrove-shrouded shallow waters of Vatoa reef. When they came out of the shelter of Vatoa, they saw that the boat was still headed toward Naniqaniqa, deceived by the leaping flames of the fire on the beach.

As they made for the wide stretch of open water that separates Vatoa from Galoa, the big boat saw them and turned to give chase. Somehow they reached the safety of the mangroves on the windward side of Galoa and hurriedly tied up the boat. Placing his feet on the protective sands of his home island, Malé's grandfather shouted into the blue moonlit night, "Here I am, come and get me, come and get me," and slid into the bush.

When they reached the village, Malé's grandfather woke all of the men and told them to get their spears to defend themselves. When the tide came up, the big boat came across the reef toward the village. Malé's grandfather had a gun; Malé doesn't know how he got it. It was the only gun on the island. When the big boat touched the sand and the men jumped ashore, Malé's grandfather fired one shot into the air at the sail. The men from Macuata ran back to the boat and paddled away. It wasn't so long ago, maybe fifty years.

Malé tells me his grandfather, Jone Varawa, was a very dangerous man. He had a face like a *tēvoro*; it was frightening to look at. He was silent most of the time, but when he talked, everyone sat quiet and listened. He had a lot of stories to tell.

We talk about old things, and I ask Malé what he knows of old things. Did he ever see anyone eaten? No, but he tells me about the death and funeral of the last chief of Galoa a few years ago. Malé was big then, and it was the job of his *mataquali* to prepare the chief's body. The chief's face was blackened—black, black, black—so that only the dead staring eyes shone out of a black face. His body was oiled so that it gleamed; and, naked, he was wrapped in a *tapa* and laid on a *tapa* in a special house built for his body. The men in Malé's family had the responsibility of guarding the body, and no one was allowed to pass, man or animal, or they would be killed, clubbed to death. Malé killed a dog that wandered over the invisible line drawn around the body of the dead chief. There were four guards at a time, one at each corner of the house. They stood rigidly, holding a specially carved hardwood club over their shoulders, arms straight out in front of them, heads not turning, no one speaking. They stood for an hour; then they were relieved by other guards. There was no wailing or mourning for the chief. The women did not cry the customary *oilei* for the dead. There was only the continuous blowing of the great conch shell (*davui*), all day and all night, day and night only the eerie hollow sounding of the great seashell, and inside the house the white staring eyes in the black, black face of the dead chief.

That is all I remember of the old things, Malé tells me. But not long ago in Lekutu, when they built the big, big *bure* with the great high roof, the one for the chief of Lekutu, they did bury a man at the corner post. It was an Indian man who came around selling Fijian tobacco. When it was time to put the post into the ground to start the *bure*, the Indian came near asking if anyone wanted to buy smoke.

"Yes, yes, come, I buy," they said, and then they grabbed the man and killed him and buried him in the hole and set the big post over his body. The *bure* is gone now; it no longer stands in the village.

The new church in Lekutu was built on the *yavu* of the old *bure*. But Malé tells me he saw the house when he was a little boy; and the post, the big, big post, was so big you could not put your arms around it. There are no more posts like that around anymore.

The *bure* was so big that full-size coconut trees were used for the rafters: thirty-six of them to a side, and the dovetailed beams lashed together with *magimagi*. Little by little the house fell apart, for there was no one left who knew how to repair a great house. It was the last great chiefly *bure* on Vanua Levu. It was his mother's family, the chiefly family of Lekutu. We are related to the chiefs, says Malé, looking into the tunnel of his own time.

We roll another cigarette, and I light the Primus to boil the water for tea. Outside the cold sky glitters. I ask Malé if the grave of the chief is nearby.

"Yes," he says, "over there," and motions toward the hill. "Do you want to go and see?" he says.

"Yes, if you come with me."

"No, you go, you go," Malé looks at me.

"If I go, something bad will happen?" I ask.

Malé looks at me. "Go, you go."

And I laugh and say no.

The night before, deep in sleep, I heard Malé moaning and whimpering and thrashing around. It was the *tēvoro*. I shook him awake and called, "Malé, Malé, *maqa tēvoro, no tēvoro*."

He jumped out of bed and ran outside to the *baka* tree, cursing and shouting and spitting; then he came back into the *bure*, still cursing, but now in English. He had chased away the *tēvoro*.

We drink our tea, and I massage Malé's back. His muscles are hard and beautifully defined. His skin gleams in the soft light. The homemade coconut oil, with the faint nutlike fragrance, makes his back shine like satin. I work at the knots and tightened muscles to knead away the soreness. Malé turns over on his back, and I massage his chest and arms. He sighs and relaxes into sleep. I turn down the lamp to a soft glow, slip in under the warm covers, and stare at the shadows, thinking of the *tabu* grave of the dead chief somewhere nearby in the silent bush.

THE GOD'S WAY

It has been more than a month since I sat at this typewriter, and in the interval much has happened. It has been a month of indecision and confusion, and it climaxed when I left the island after a very bad scene with Malé's father. We had gone to Lekutu with Sione, Vili, Qare, Ima, Una, and Va for a sports tournament. The boys played rugby, and the girls played net ball, which is a grass-court basketball. There was no way I could possibly participate; I had lost more than twenty-five pounds and was weak. I looked, and felt . . . old. It was a boring, unhappy time for me, feeling very much out of it, watching everyone else friendly and active: the children following me around staring at me, the men drinking grog and hugging one another and holding hands, the girls giggling and flirting with the young boys. I wandered around holding onto my pride as best I could, yet not being very successful about it.

Late in the afternoon the tide was going out, and it was time to go, for if we missed the tide we would be stuck in Lekutu all night. It took almost two hours to round up everyone, and Ima and Va still couldn't be found, so I said that we would go without them. Malé said no; it wasn't *va Viti* (the Fijian way), but I didn't care. I had had enough of *va Viti*, and I said we go.

We left in a miserable silence with Sione, Malé's older brother, cousins Qare and Seine, and Una, the only sister we could find. We arrived almost at dark, and Malé's father received us with his usual good cheer. We ate and went to sleep. In the middle of the night, I was awakened by Malé's father, who was crouching in our doorway like an enraged lion and screaming at me, "Why you no bring Va? Why you no bring Va?" beating a stick on the mat as if he would like to beat it on me. It was the first time he had allowed his native temper to shine out at me, and I was furious. The substance of the rage was that we had gone off and left the girls and brought the cousins instead. Momo had waited until the last boat from Galoa came back from Lekutu, and Ima and Va still had not come home.

The longer he crouched there in the darkness and beat the stick, the more I resolved that Fiji and a Fijian family were not for me. He went away for a few moments and returned with our big flashlight, shining it in my face as I sat in bed, and screamed at me. I held in my anger and said nothing, fearing that the scene would turn really ugly if I got mad too.

After he left I told Malé that no one in my life had ever treated me like that, which was, discounting fights with lovers, the truth. Moreover, I was furious because the rage centered on the girls, and nothing was said of Vili, who had left a wife and two children home in the *koro* and had stayed in Lekutu to drink grog with the boys all night. My feminist fury surfaced, for it was by my standards manifestly unfair to not allow the girls one day of fun when the boys were allowed to do almost anything they liked. To me it was another instance of the overbearing dominance of the male, and I could no longer cope with it.

We sat there in an uneasy silence, drinking coffee and smoking cigarettes, and finally fell into a distant and unloving sleep. I can't remember if I dreamed, but I woke early, full of resolve to leave the island, Malé or no Malé.

To my surprise Malé told me to take some *yaqona* to his father and to tell him that no one had behaved like that to me before and that Malé and I were leaving and were not sure if we would return. "That way my father will know what he did to you."

It was Sunday morning, so I bathed, brushed my teeth, dressed in a clean *sulu*, and waited for Momo to come back from church. I tried to calm and center myself, for the best intentions go astray in a foreign language, and I knew I had to be pretty clear about what I wanted, or it would get lost in the rhetoric. When the family had finished lunch, I went into the *bure* and told Momo that I wanted to talk with him and Sione, who could act as an interpreter, as soon as they could meet me in Botoi. One helpful thing about Fijian custom is the readiness to meet and discuss problems, and though often they are not fully resolved, the process itself is restorative.

I left Malé, who had retreated into a semistupor as his defense against family conflict, and went to see his father, carrying the newspaper-wrapped bundle of *yaqona* roots. We sat on the mat in Botoi—Momo, Sione, and I—and I presented the *yaqona* and said I had come to talk to him. Momo asked why I had brought *yaqona*; it was Sunday and not customary to present *yaqona* on Sunday. Was it a *soro* (an offering of forgiving)? No, it was just to let him know that what I had to say was very serious.

The new mat gleamed in a spray of sunshine that entered from the open doorway. From the church I could hear the sweet harmonious voices of Sunday hymns. I looked up at the high-pitched roof, at the elaborate pattern of bamboo strips and vine lashings, at the still-sweet-smelling grass that we had cut on Vedrala. I looked past Momo at the great *tapa* hanging on the far wall; its beautiful black-and-white-and-brown design imprinted in my mind all that I loved of Fiji and this village. Momo's face was tense, for he knew what was coming, Sione's an impassive mask of no expression.

I began. "I have always tried to be good to this family. Whatever they have wanted or needed I have tried to give. When you needed a new outboard, I got it. We are having a new boat built to help the family; I have given money and food. I have done all that I could, and no one—not my father or mother or my husband—has ever treated me the way you did, Momo, last night when you woke me from a sound sleep, screaming at me and beating the stick, waving the light in my face. No one has ever done that, and so I am leaving the village, and Malé is going with me, and we do not know if we will ever return. There is no sense in going to church and praying and reading the Bible and saying grace if there is no *loloma* (the love that is inside), the kind forgiving love that Jesus died to teach us. All that is empty, nothing, if there is no love. Last night you took the stick and beat it at me, and beat Una."

I embroidered and embellished the story, the meaning, hoping to get his attention, hoping to make him undersand that he could not treat women like that, that he had no right to punish the girls, who worked hard, and let the boys do whatever they wished. When everything I could think of had been said, Momo asked me if I was finished, and I said yes.

"Thank you for the *yaqona*," and he patted the newspaper-wrapped bundle. "Bringing it shows you have respect for *va Viti*. When all of the boats had returned from Lekutu late last night, and Va and Ima were not back, I was very angry. You should have brought Va and Ima with you and not Qare and Seine. They are only cousins. Ima went off to another village with a boy. We cannot let our young girls do this. I did not mean to beat the stick at you; it was for Una, who had disobeyed me. I told her to stay home last night, and she went around the *koro*. Even the God gets angry. It says in the Bible that the God was so angry with his people for not obeying him that he killed everybody and destroyed the towns and all of the animals. Sometimes I have to use the stick to teach the girls. I am sorry that I did that to you; I was very angry. Even the God forgives. If you go, I cannot stop you. I do not want you to go. If Malé goes, I cannot stop him. I do not want him to go, and if he does I will treat him like the prodigal son in the Bible. I hope you will not go."

When he had finished, he patted the *yaqona*, said the ceremonial words of thanks, clapped his cupped hands in the *cobo*, the sound soothing the stillness, and looked at me. I asked if I might say something else.

"The message of the Bible is love. It is the reason that Christ died on the cross, to teach us to love one another. That means to love your daughters as you do your sons. Man is not the God, God is the God. It is up to Malé if he wants to go. I am going. I do not know if I am coming back." I clapped my cupped hands three times, said thank you, and left the *bure*.

A bird sang from a breadfruit tree; a rooster crowed. The

The village on Galoa.

Malé and his father.

Dressed for a *meke* on Galoa. Momo is second from right in the front row.

Malé and cousin Mone rebuilding Mone's *bure*.

Malé with the pigs on their way to the well, Vedrala.

The *bure* on Vedrala.

The author at home on Vedrala.

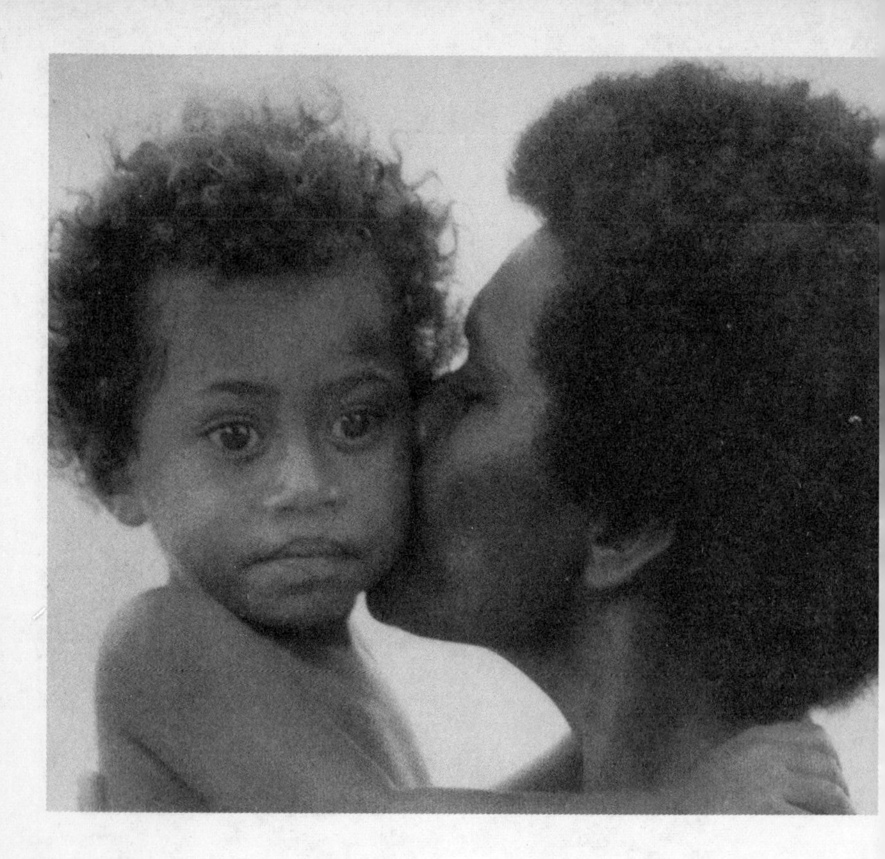

Vaseva and small Joana.

yellow leaves of the ginger gleamed in the sunshine, its fragrant flower head heavy with blossoms. As I walked down the clean-swept path toward our *bure*, I heard the incoming tide announcing its intentions on the sand. The upended shell of the *vonu* had been cleaned of its message of dried blood. All was still and uncompromised.

I went into our house, the house I was beginning to love, and found Malé asleep in a fetal ball, his face turned into the pillows, the radio next to his ear broadcasting a Sunday sermon. Pulled violently in opposite directions, he could only sleep. He woke when he heard me and asked what his father had said. I told him as simply as I could, trying to be fair, and waited.

"Will you go with me?"

"Yes. We go."

In the morning I packed, for what journey I did not know, and Malé sat outside talking to his cousin. We invited Toa inside for coffee and biscuits, and the three of us sat on the mat. Toa shook his head and looked at us, "*Isa*, Joana, *isa*."

Una came into the *bure* and said that Malé's father wanted to talk with him before he went. I considered it a way of delaying us until the bus had gone, and took my things outside and put them on the punt. I asked Una if her father had beaten her the night before. She said he had called her over in the darkness and hit her once with the stick; then she had run away, and he couldn't find her. She had slept at her aunt's house, and now all was calm. Una seemed unperturbed; for her the whole business was over. She said that after Vili took Malé and me to meet the bus in Nakadrudru, he was going to take the boat to Naivaka to bring back Ima. I asked if her father would beat Ima when she came back.

"No, my brothers will."

I told Una to tell her father that if I heard that he or the brothers had beaten Ima that it was all over, *sa oti*, finished forever. I pulled my things into the boat, getting my *sulu* soaked in the high tide, and told Una to go tell Malé that I was leaving right now, that if he wanted to come with me, come now; I'm going. I was settled in the boat, trying to lower the heavy outboard into the water, when Malé came down the path.

He climbed into the boat and said, "My father doesn't want me to go for long. He is sick and wants me to come home soon so we can chase the *vonu* so he can take it to Suva. My father has *mana*. If I go away when he is angry with me, something bad can happen to me. Something on the bus. We don't know."

I asked if Momo was angry. "No," said Malé, "he blessed us."

As we sat in the boat waiting for Vili, I realized that all of my words, my intentions, were sand tossed in the tides of this culture. I said nothing. I was not angry with Malé; what I was asking of him was too hard. He had obeyed his father and asked me to marry him. It was not his choice, but his father's desire to improve the situation of the family, that had inspired his courtship. Now I was asking him to disobey his father if he loved me. There was no way he could.

Slowly we pulled away from Galoa, and I looked at the shoreline, the tall arching palms, the dense shiny foliage of the mangroves, the grass-covered hills turning golden in the winter sun. I loved this island and knew that I would be coming back, that I had not finished my life here. I looked back at Malé driving the boat and flicked my tongue at him. He smiled and flicked his tongue back at me. This is the day we are given. There is no sense in spoiling it with tears.

As we were getting on the bus to go to Labasa, Vili said

something to Malé, who turned to me and asked if we could come back on Thursday. It was Monday. I said, "No, we come back on Friday." It was settled.

* * *

We stayed away two weeks. When we got to Labasa to renew my visa, I was told that it was my last month in the country. I could stay only six months on a visitor's visa. I would have to go to Suva to see what could be done at the immigration office. We went to Savusavu for a three-day holiday, a luxury of hot showers, cold beer, room service, and privacy. At two-thirty in the morning, we boarded a big passenger ferry for the long trip to Suva.

On board the *Ovalau* the videotape ran uninterrupted for fourteen hours—an orgy of boxing, Kung Fu, female body-building, and a long extravagant tribute, in sequined splendor, to Martin Luther King, Jr., amid the rolling heavy shallow seas of Fiji. Watching the video and looking back at the care-worn impassive faces of the Fijian women as they gazed at nearly naked muscle-bound girls posing and posturing and flashing false smiles at the camera, I wondered what was in the future for Fiji, a Fiji just barely out of cannibalism and now bound, it seemed, for the psychic cannibalism of modern life. The glittery expensive costumes of the black performers contrasted with the stark black-and-white footage of a South laboring to free itself, and the confident voice of Martin Luther King's dream seemed drowned in the overproduced music of his tribute. I tried to explain to Malé what it was all about. He was not interested. Outside, green islands floated past the dirty windows of the boat, islands of slow villages and cousins strolling hand in hand, repeating endlessly the stories—of fishing, of chasing the *vonu*—

villages of indolence and the drifting gray-blue smoke of cooking fires. Inside, the blurred garish colors of human caricatures bobbed and grimaced on the small screen.

The voice of the bored Indian clerk in the immigration office spoke with the accents of God. Leave Fiji or marry. True, I could leave Fiji for a few days; it was necessary only that I reenter from a foreign destination, but the nearest place was Tonga. If I was going to Tonga, there was no sense in going for a weekend, and to stay would be expensive. Besides, I had no desire to leave Fiji; my life with Malé and the family was engrossing and complex enough to keep my interest. It was good to be somewhere where I was considered part of the family—in spite of all of the difficulties. I had little desire to experience the discomfort of being the stranger, and if I went to Tonga it would be like starting all over again. The truth was that I had no desire to go anywhere; my problems with Malé bound me to him as much as our possibilities did. Leave or marry? I asked Malé what he wanted.

"It's up to you."

"Do you want to get married?"

"I don't know—if you want to."

He was much the same Malé, more interested in whether we were going to drink grog with his uncle that night than in whether we would get married. Curiously enough his very indifference seemed relaxing. It wasn't any big deal. If we got married, we would be married; if not, then not. Marriage in Fiji is not the climactic romantic moment it is in the United States. It is another thread in the fabric of life, an important one, but there are others of equal importance. It is the difference between being tribal and being isolated from communal responsibilities and support. Malé's indifference was probably not personal; it seemed likely that it was truly okay one way or the other. Not much would change; we would still be the same people living in much the same way.

What then for me? Should I marry Malé? All of the things that spoke against it also spoke for it. In some curious way our great cultural differences balanced the scale. We were vastly different; perhaps we might never really understand each other, yet had I ever really understood any other person? It was just easier to assume understanding someone whose cultural background was the same. Being with Malé was intense and demanding. It required my full attention. It was appealing because it was hard.

Malé had been raised to dominate, to consider himself right all of the time. I had been raised to consider men wrong all of the time. My mother is a fighter. She taught me her way. Malé and I would fight, it was inevitable. I had never been attracted to anyone with whom I couldn't fight; it was my inheritance, as much a part of me as my green eyes or bushy hair. Marriage with anyone is a tenuous shaky business. It is something that has to be built, experience by experience, into a stable structure.

And there was, underneath all of my reflection and reasoning, a curious feeling of being led, a feeling that my life was not a car I was driving down a highway, but a path that I followed to an unknown destination. My whole relationship with Malé, and with Momo, with the village, made me feel that something was being offered to me to take whole, that my task was to learn to accept with grace what was offered, to follow the Tao.

We went to the government office and for thirteen dollars took out a License of Intention to Marry. As we filled out the papers, I looked at the blue boxes stacked behind the counter: thousands of forms neatly categorized—birth, marriage, death, Fijian, Indian, other. We would add our paper to the pile; ultimately it would be filed in one of the boxes. Maleli Rakanace

Varawa, fisherman-farmer, and Joan Bobrin McIntyre, writer, were to be wed. Birthplace Galoa Island, Bua; and birthplace Los Angeles, California, were to come together and say yes. Yes. I will stay and obey. Yes. I will love. Yes. I will make this life good for both of us. Because that is what is offered. Because that is what God and fate are calling for.

We went home and told the news to Malé's father, thin and worried when we arrived, ecstatic when he heard of our decision. What he had hoped and prayed for had occurred. He pointed skyward and said, "*Au sa marau, au sa marau* (I am happy, I am happy). It's the God!"

That evening we sat on the floor of the eating house, the house in which I had shared my first meal with the family the first day I was on the island; and we planned the wedding. One cow, two pigs, fifty bundles of *dalo*, flour, sugar, rice, oil, tea, curry, onions, salt, *yaqona*, cigarattes—enough to feed all who would come: the relatives from Lekutu and Tavea, the whole village of Galoa. The time was set by the necessities of the visa; two weeks were all that I had left, so in two weeks, on Saturday, June 7, we would wed. I wished for a full-moon wedding; now I could only hope for a new moon. As soon as it was settled, I felt calm and happy. Something deeper than romantic love had been resolved. I was following the way.

WEDDING

A heavy gray unseasonal rain marked our wedding day, a bizarre day through which I moved as if in a dream. I was a strange distant person, watching the celebration from afar. The rain lent a peculiar urgency to what would have been more ordinary, the preparations of the family and *mataquali* for a *magiti* (feast) that was the central fact of importance. I recalled reading about a young Fijian girl of high status who was given to an old chief as a bride; and I had the sense of her as an object of little importance, a prop in the transaction which centered and displayed the power of the old men. On the day of our wedding, I felt the same sense of removal of self and passion.

The night before we were to marry, Malé, cousins Penioni and Kava, and I had gone fishing, a kind of men's party at which I was the spectator. The reason for the fishing trip was to catch fish for the next day's feasting, but the three men got drunk, very, very drunk, finishing a full bottle of rum between them in less than an hour. We were anchored off Vatoa at the bikini, a shallow-water marker, and not catching any fish. Penioni passed out, half-unconscious on the cabin floor. Kava talked incessantly of how good it was for me to marry Malé, for I would give him everything he wanted. Malé got violently happy and careless, and I found myself more and more angry and distant.

By the time we returned to the island at midnight, I was in no mood for marriage or Malé. He called me to get off the boat and come ashore; I wanted to stay and cool my temper and said I would follow in a little while. Malé got really ugly and threatened me, grabbing my shirt and pulling me off the boat, his face flashing fury in the wavering light of the flashlight. Kava tried to calm him, telling me to come along, and I held my anger, knowing that if I lost control, Malé might really hurt me. I went along, outwardly calm but furious inside, and followed Malé to Botoi, where he went inside to drink grog with the family.

I sat outside in the soft spangled darkness and talked with the still-drunk Kava, who kept repeating how good it would be for me to marry Malé, for I was rich and could buy anything he wanted. The more Kava talked, the less I felt like marrying Malé. A few minutes later Malé came outside to piss, and I stopped him to confront his anger. Our words, like all words of conflict, are lost now, but Malé drunk is dangerous, inaccessible, and overcome with violence. He did not really hurt me, but showed that he could and would if I did not submit; and pushing me away, he cursed me and went back into the *bure*.

Momo came out and told me to forget it; Malé was drunk, and it would be fine in the morning. Just let him alone, he said. Marika, one of Malé's cousins, came to sit next to me, telling me that all Fijian men acted that way; it was nothing. Just forget it, forget it.

I went into Botoi, sitting far away from Malé and not looking at him. I looked into the faces of the women, the cousins and aunts, for an answer. Their faces were compassionate and distant. Forget it, they seemed to say. This is the life we are given. Everyone joked and drank *bilo* after *bilo* of *yaqona*. I felt calm, from the *qona*, with the peculiar calm that only danger grants, and drank and joked with Qare and Marika, and accepted cigarettes gracefully from Uncle Malé, who was being especially kind, and resolved that I would leave in the morning.

Late that night we lay in the same bed, back to back, strangers on the eve of marriage. Malé left at dawn with his father to go to the mainland to pick up the pig and cow that were to be sacrificed for our feast, and I looked at the littered *bure* and listlessly tried to pack my clothes.

Everything I touched or looked at was swollen with sadness; my possessions seemed to mock me. Everything I could love was here, and yet my pride told me that I must leave. *Isa*, said the *tapa*; *isa*, said the *tabua*, the woven walls, the disordered table. I wanted to go away, yet where was I to go, and how would I get there? I could not even navigate the river by myself, and if I did, where could I leave the boat? *Tapa*, *tabua*, bright fringed mats, books, and ukuleles—possessions held me here. Then came a gray wall of rain blotting out the hazy ridges of the mainland, obscuring the sky and sea, a dense dark rain that turned every-

thing into one wall of water. I was forced to sit and watch my mind go around, comforted and slowed by the rain, which said forget it, forget it for now, for there is nowhere to go and no way to go.

So I sat in the *bure*, calm, with a grace that was given to me by the violence of man and the violence of wind and water, and waited for the far-off sound, the high whining that would tell me that Malé and his father were returning.

Anger is a bond. I did not want to talk to Malé; I did not want to be with him, but I wanted him near, and I wanted him safe. Finally the boat came to rest far offshore; the tide was way out, and I watched through the doorway as the men, black silhouettes in the gray driving rain, unloaded the grotesque mutilated shape of what a few hours ago had been a living cow—legs and haunches, slabs of blood-dripping meat dangled and dragged—the tail swishing lazily in the shallow dirty water of the reef. Slowly they came near—Kava, Toa, Vili, Momo, Malé—carrying their tokens for the love feast.

Then came the pig bound to the bamboo pole, hanging helplessly, swollen belly, tail twisting and turning. The huge pig, with fear-driven eyes, was laid under the *baka* tree at the door of the house of the *tēvoro*. Bright green piles of freshly cut fern, emerald green leaves of *dalo*, huge bundles of *dalo* roots bound by the stalks, great piles of cassava and yams in newly woven coconut-leaf baskets glittered and sang of death in the gray rain. Food, food, food. Everyone was intensely happy and excited. There was plenty of food for the feast. Wet *sulu* clung to the wet shivering bodies as the women and men, the aunts and uncles and cousins, worked in the pelting cold rain, laughing and calling, to prepare the *magiti*. When they cut open the cow, she had an unborn calf inside; and the pig too contained babies—five of them. I watched and listened as a stranger, in awe of the omens.

From time to time Malé would pass, a questioning look in his eyes. Hello. Hello. Flat voice, no sign in answer. He was through, the anger gone with the alcohol, but we were strangers.

Late in the afternoon, after the cow had been cut up into bite-size pieces and was simmering in the big stew pots, and the pig was steaming in the *lovo* along with the *dalo* and yams and cassava, Malé came into the *bure* and sat on the mat looking at me. He rolled a cigarette, moved slowly, quietly, "I'm sorry for what I did last night. I was drunk. When I'm drunk, you must do what I say, or I get very angry."

"No. It's not like that. You cannot control the world with anger. You must control yourself. Last night I had to control myself. I was angry too. If I had let my anger out, you could have killed me. Then today, instead of a feast, there would be a funeral, and you would go to prison for the rest of your life. For what? There would be the *oilei*, not the sound of laughter; and the cow and pig would be eaten, not at a wedding, but at a funeral. For what? Because I did not do as you said. No good, Malé, no good."

"I promise it will never happen again. I'm sorry."

The short, terse sentences dropped between us, sentences formed simply because we did not share a language of rhetoric, of deception and elaboration. Good, no good. Simple, restrained. And while we talked, I knew that I had forgiven him already. Forgiven because it was the only thing I could do. Life was carrying us along in its current; we went either willingly or fighting, but we had to go with it. God had sent me to Fiji to marry Malé; God had returned Malé to the *koro* from Suva just two days before I came to the island. It was all planned somewhere else. Did I want to marry? I wasn't sure, but I didn't want to leave either; and there was the cow and pig and *dalo*, and all of the people; and Momo had just left for Lekutu to get the minis-

ter. I had the choice between chaos or continuing. I went along like the virgin bride in the book, following Malé, to the big *bure* to dress for our wedding.

In the dim gray late-afternoon light, Botoi was lit by a single Coleman gas lamp hanging from the crossbeam. The first thing I saw on entering was Malé's aunt Mela sitting on the mat shaving grains of sandalwood from a stick of heartwood, the sawdust from the sandalwood perfuming the air. Sandalwood, a scent I have always loved, bought in distant places, imprisoned in small vials of oil, put under my arms to mingle my own smell with the life smell of the tree, a perfect fragrance, wood, flower, smoke. I remembered the last night I had spent on Galoa the time I was leaving for Hawaii last year. Malé had taken my hand in a gentle courtship to lead me across the village to see something he thought I might like: in the same yellow-white light of the gas lamp, a number of men and women, at that time nameless, were preparing a sandalwood tree for market, shaving the outer wood away from the close-grained oily red heartwood, gracing the night air with the scent of sandalwood and the sound of laughter. Something then he wanted to share—and now—sandalwood heart grains piled on the mat, shining golden in the golden light.

Dressing was a blur. I had bathed earlier in the day under a pouring spout of rainwater cascading off the iron roof of the eating house, and dressed in a simple *tapa*-printed cotton *sulu* and *jaba*. I was clean and unadorned. Aunt Sereima wrapped a floor-length piece of tapa around my waist, tying it with a string to form a great billowing skirt of patterned brown and black and white; then she put another shorter piece over that, creating a double skirt of astonishing beauty. She slathered homemade coconut oil, smelling of the nut and flowers, on my arms and legs and face, and put a *salusalu* around my neck—streamers of white *tapa* woven with purple bougainvillea blossoms and ferns—and sprinkled sandalwood grains on my hair. I looked in the mirror—small head, wet clinging hair, distant face—and turned to look at Malé.

A dark prince in a pale blue shirt and a floor-length *tapa* skirt, his dense black hair glittered with raindrops and the golden swath of sandalwood. He looked happy and excited; he did not catch my eye; nothing passed between us. Monomono, my friend, who would stand with me, and Kava, Malé's cousin, were also dressed in *tapa* and anointed. There was no light to take pictures, but we all posed for one anyway, standing stiffly under the Coleman lamp.

We could not sit down because of the skirts, and stood in the gathering darkness waiting to go to the *Lotu* (Christian church). The aunties chattered and laughed. Monomono spoke to me in English. The minister would say the vows in Fijian, I was to copy his voice and words; she would ask him to speak slowly. The woven ceiling of the *bure* arched above us into the shadows; the lamp cast its soft mellow light on the *tapa*. I gave Malé the plain gold wedding ring I had bought in Labasa. He put it into the pocket of his shirt.

"This ring," I said, "is important to me."

When it was time to go, we all lined up under the heavy green tarpaulin I had bought for the boat, to shelter me from the impossible Fijian sun while fishing, and formed a procession to go to the church. Then it was funny—standing under the tarp in the doorway of the great *bure*, holding it aloft over our scented heads in the driving rain, filing into the dimly lit church, to marry.

We stood on the new beautifully woven mat that Nei had made especially for the wedding. Kava, Malé, Monomono, and I listened to the words rolling into the darkness, strange words that would alter the course of the rest of my life. I looked up at the stained-glass window at the last light of day, gray twilight floating in, turned blue and green and lavender, at the eternal dove flying in an eternal fantasy sky of colored glass and light. As the minister spoke the vows, Malé repeated them in a strong, sure voice, a voice that was not lying or deceptive. He was speaking truly, out of his heart, out of the center of his strength and assurance. The gold ring gleamed in the dark hand of the minister as he held it over the Bible and blessed it, and blessed our marriage. I looked at the ring and felt the strength of Malé's hand as he held mine. We were married.

⯭ ⯭ ⯭

At the close of the ceremony, we untied the *tapa* skirts and heaped them onto the mat, folded the mat over the *tapa*, and presented the bundle to the minister, and went forward to sign the marriage license, a plain printed government form that would take its place in a file of vital statistics.

From out of the gloom behind us came a procession of people—relatives and friends—to sniff our faces and shake our hands, to wish us life and health and God's blessing. *Bula, bula re, kalougata.*

The rain had stopped, leaving a clear glittering sky and the fresh smells of rain-washed earth. We went back to Botoi, laughing and relaxed, to wait for the setting of the wedding feast.

Momo was almost hysterical with joy. I could hear his voice outside ordering the arrangements—*totolo, totolo* (hurry, hurry). The feast was essential. It was not symbolic, but the actual sharing of our wealth with the family and the village. Giving and sharing food, the source of life, was our demonstration of our love for the family, for the tribe. It was vital that there be plenty to eat, that we give fully and freely.

The feast had been set up in the old church. A long table stood in the center, covered with clean tablecloths and set with dishes heaped with curried cow meat, pig meat, turtle, fish, *dalo* leaves in coconut cream, cassava, *dalo*, yams. Benches had been set up along the walls, and we took our place at the far end opposite the head of the table. When everyone who could get inside was seated, Aunt Mili took a knife and cut into a huge snow white yam. This was our wedding cake, and she cut a piece of the yam and fed it to Malé, who then received another piece and fed it to me. The family feeds the family. The lineage continues.

Then everyone ate. Eat, eat, eat some more. Eat plenty. I chatted with the minister from Lekutu, who spoke English perfectly, and then with Malé's Uncle Navi, who had made a touching speech during the ceremony in the *Lotu*, a speech in English, for my benefit. He had said that this was indeed a historic occasion, one that would be remembered and watched, for it was truly the first time that a European and a Fijian had come together in marriage in the history of this village. It was up to us. I thanked Uncle Navi for his thoughtfulness. He had drawn me into the circle with words that I could understand and relate to. It was very kind.

I looked at Momo, at his intense energy and pride; at Nei, quietly hopeful; at the girls, Ima, Suli, Una, Va, serving the food, filling the dishes over and over from the big pots that were clustered in the back of the church. Banoko smiled shlyly at me; Vili gleamed. This was my family now, and I was one of them. When everyone had eaten their fill, we went back into Botoi to await the clearing of the church and the setting up of the *yaqona*.

"Come, Joana, we go drink grog," and Malé took my hand and led me back into the old church. Huge mats had been laid on the bare wooden floor; a giant *tanoa*, the biggest in the village, was filled with *yaqona*. Malé sat in the center behind the *tanoa*, Vili and Qare on either side. I sat between Momo and the minister.

"*Talo yaqona va turaga*," called Momo, and *bilo* after *bilo* was drawn out of the great bowl and handed around the circle. The guitars and ukuleles were brought in, and Malé, Qare, and Vili sang song after song. Their voices blended so perfectly that it was impossible to separate them. There was plenty of grog, plenty of food, and the party lasted for three days. It was the first marriage performed in the new church, and the first feast given in the old church. Long afterward people spoke of the food, of the cow meat and pig meat, of the *dalo* and yams.

THE DOLPHIN

Early Monday morning Malé woke me with great gentleness, "*Yadra, marama* (wake up, lady), this is the day we go to get our boat," and kissed me on the lips, smiling. We discussed the day's plan with Momo over tea, small Malé circling in the background eyeing the peanut butter and biscuits. We would go first to Tavea to pick up the men who would help carry the boat down to the water from Bill's place; then we would leave Momo on Tavea to drink *yaqona* with the helpers while Malé and I took our new boat on her maiden voyage up the Lekutu River to Nakadrudru to buy some more *yaqona* for the evening's celebration.

A good boat was vital to our life on the island, our transportation, the means of our livelihood. For a fishing family to be without a good boat was true poverty. For months we had been using Bill's old runabout, until he could complete our new boat. The family's old punt, the *Tilaro*, named appropriately for the crab, was leaking and feeble. When Malé and I had first negotiated the terms of our mutual aid, I had promised a new boat and engine so that the family could become more self-sufficient.

For years, while sending the children to school and paying their school fees, Momo had set out each afternoon to paddle to the fishing grounds to lay the net, tend it all night, and paddle back to the village for a quick cup of tea, and then on to Lekutu to sell the fish. He had done it faithfully, educating all of his children, even sending Sione and Luisa on to further schooling to become teachers, but he was tired now, and no longer capable of the eighteen- to twenty-hour days that paddling to the fishing grounds entailed.

Early this year we had gone to Labasa and bought a new 40-horsepower outboard and the lumber for the new boat. The weeks had stretched out to months, and now finally the boat was completed, ready to be joined to the outboard and a new remote-control steering system, to become the Rolls Royce of the village boats. Malé was ecstatic; his boat would be the fastest and best in the village. Momo was vastly relieved; the boat was a great step in putting the family on its feet.

Vili transferred the outboard to the old faithful *Tilaro*; we loaded the cables, steering wheel, and controls for the remote system into the punt; and Vili, Momo, Malé, and I set out for Tavea to pick up the helpers. We would get our lifting help from the uncles and cousins on Tavea because it was much closer to the mainland. There were always, no matter what the task, strong men who would cooperate in heavy labor—lifting and carrying whatever was needed—for some *yaqona* and cigarettes. The *yaqona* wasn't payment. It was to solidify the bond, to continuously reenact relationships of calm loving cooperation and respect. Working and drinking *qona* together were inseparable.

On Tavea we picked up fourteen men—cousins, uncles, and in-laws—and Momo and Vili towed the Tavea punt with the *Tilaro*. Malé and I sat in the stern of the Tavea punt, and I looked at the men crowded shoulder to shoulder in front of us. They looked fierce and dangerous, bare-chested, wearing *sulu* and tattered headbands, their muscular arms and chests covered with black homemade tattoos proclaiming sex and the ability to endure pain. It is likely that their great-grandfathers ate people; yet they were courteous and kind to me, smiling shyly in my direction. The constant drinking of *yaqona* has much to do with mellowing and relaxing these strong men, with preserving their even temper. I shudder to think what Fiji would be like if its national drink were alcohol.

We drew into the shallow muddy mangrove passage to Bill's landing and tied up the boats. A small clear stream of fresh water washed the stones of the shore, a luxury of fresh running water that tempted me to stay behind for a few minutes to bathe. Bill had rigged an old icebox as a makeshift bathtub, and I soaked happily, listening to the faint stirring sounds of a hot afternoon and the shouts and laughter coming from the house above. I climbed up a steep incline, playing with the short feathery leaves of a ground-hugging plant that closed at my touch, opening slowly when the disturbance was past: *pua hilahila* (bashful flower) in Hawaiian. I have no idea what the Fijian name is for this botanical wonder. On reaching the house I found all the men admiring the boat.

She was truly lovely, twenty feet long, eight feet wide, built of plywood covered with fiberglass. Bill had painted her hull white, the deck and cabin a soft yellow—easy on the eyes. She was touched with trim of dark blue and brilliant golden yellow and looked fast and clean. I took pictures of everyone standing around the new boat. Lifting her up as though she were of no weight, they carried her down and gently set her in the water.

Before we came, Malé had told me he was going to name our new boat *Dolphin*. It was a name full of meaning for me, for I had spent years of my life involved in the politics of saving whales and dolphins, something Malé had no knowledge of, and his choice of the name seemed like another confirmation that I was choosing the right way for myself: *Dolphin*, fitting another piece of the puzzle into place.

The men paddled back to Tavea in their punt with Momo to drink *yaqona*; Malé, Bill, and I spent an hour or so fitting the steering wheel, gas and shifting cables, and control unit into our new boat. I stayed on the shore while Bill and Malé drove the *Dolphin* around testing the controls. She looked wonderful and was worth the frustrations and confusions of the months she was under construction, for nothing involving things in Fiji is easy. There had been weeks and weeks of delay in getting the right materials, days of anger and resentment and misunderstanding, but she was finished now, and all our lives would be easier.

"Bye, Bill, bye. Thanks for the boat. It's really nice. Bye." And Malé sped off like a demon, anxious to test the *Dolphin* to the limit. We carved great circles of bubbly white in the clear blue-green water above the reef, flew around in a dazzle of sparkling spray. Malé had his new boat. He was king!

We headed around the big shoal that borders the mouth of the Lekutu River and sped up the river, leaving a rooster-tail plume to mark our passage. I had traveled this river in many moods since the first time I rode its sinuous curves to the sea, seen it through a mist of tears, or not seen it at all in anger. At times its cool green beauty was dotted with the floating silk blossoms of the *vaivai*. I had watched the gray herons expectant on the banks, and one night in supreme excitement seen its dark waters flashing with thousands of fish who had come up the river to spawn. But I could never know it in the intimate way that Malé knew its shallows and eddies, every rock, every bend and curve, familiar and certain.

Somewhere, hidden now in the encroaching mangroves, was

the site of the first settlement of Galoa, the historic village that was later moved to the island. Uncle Paula had told me about the village, which once guarded the old village of Lekutu from attack from the sea. I wondered what clues lay scattered there in the bush, what memories prowled unseen.

That night, after we had finished dinner, Momo called Malé and me outside to sit before the eating house and receive the offerings of the village women for the new boat. *Cicicere*: the ceremony of happiness for the new boat. The women come in the darkness bringing great bundles of *voivoi*, piles of clean used clothing, *sulu*. They carry the unrolled dry leaves in big tied bunches above their heads, pile them and the *sulu* on the ground in front of us, and sit down in a semicircle facing us. Malé, Momo, Nei, Vili, Imi, Una, Va, and I sit on the other side of the heaped offerings. Momo speaks of his happiness at the new boat, of his thanks to the God, prays for life and health, *me ua ni dua leqa* (that there not be any trouble). When he is finished, he presents a bundle of *yaqona* to the women sitting gracefully on the ground. They *cobo* in response; and one, whom I cannot identify in the darkness, murmurs a soft speech of thanks. The voices of the women sound soothing and sweet. Five packages of cigarettes are given, to be shared among the women, and then they all file past, kissing me in the Fijian fashion, shaking my hand, and kissing Malé. As they pass, I recognize them: Aunty Amele, Aunty Sovuqa, the pastor's wife, Seine, Marselina, Sala, Aunty Una. Aunty Sovuqa has presented us with a *tabua*. It is my third tooth: the first I received on Christmas Eve; the second when Momo returned from Suva, as appreciation for helping him get the *vonu*; and now the third, for the new boat. The *Dolphin* receives the *tabua*; the new plywood runabout is honored in much the same way the ancient canoe was honored, with the things of value, the Fijian things. In earlier days the *sulu* and clothing were probably *tapa*, the *voivoi* fashioned into mats, but the sense prevails, and the *tabua* remains unchanged. It is the precious gift of the Fijian, not to be held and hoarded, but to be passed on in another gift giving, for it is the ceremony of giving, the ancient words repeated into the darkness, that is the essential meaning.

The boat is our life. She will carry and care for us. She will feed us and others. She has been blessed.

SIGA TABU

The faded red *sulu* that curtains the outhouse door proclaims "Follow the Sun to Fiji" and depicts cartoon islands marked with thick yellow lines for roads and thick blue lines for coastlines. The roads are few, and many end abruptly. There are the two big islands—Viti Levu and Vanua Levu—and a number of little blobs for the smaller ones. The *sulu* flaps lazily on its fishline curtain rod; a stick sewn into the bottom keeps it from disclosing whoever might be sitting on the cement toilet set above the two 40-gallon drums that comprise the facility. The outhouse is made of coconut matting with a coconut-mat roof and is somewhat cool. It is located in a little cluster of similar outhouses and bathing houses just next to the pigpens on a spur of sand called the *muni savusavu*. The proximity of the pigpens lends interest to the outhouses, for there I can sit in indolent solitude and watch the coming and going of the pigs. The pens enclose the big ones, who if left to their own would head into the bush and wreak havoc on the cassava gardens. The nursing babies squeeze in and out under the wire and often burrow their noses in the cool sand just outside the door of the outhouse. If not startled, it is a charming companionship, but at the slightest noise from inside the red curtain, they scamper for the safety of someplace else.

Toa and Sala's big white pig had nine babies three months

ago; and they have grown into a rooting, grunting, pushing brood, always together on the march for food. When first born, they looked much like silky white puppies, almost blind, with pink noses, playing with one another and sleeping in the sun. Now they look like storybook pigs—fat-jowled and short-legged—and leave great wet furrows in the beach, where they dig for unseen delicacies in the sand. They have recently been joined by Monomono's two brown and two black baby pigs; and once I had the ridiculous pleasure of watching one of Toa's big white adolescent pigs nursing beside the two tiny black ones at the swinging teats of Monomono's pig mother. I think if left alone the pigs would soon comprise a tribe of mixed colors and ages, all trotting along together, a vast free-ranging herd. But, alas, their days of sleeping next to the outhouse are numbered, and one of these days they will go into the *lovo* as part of a village or funeral feast.

Today is *Siga Tabu* (Sunday). It seems to be the day for watching pigs and chickens and the bronze glitter of lizards, and taking long naps in the afternoon, caressed by the hymns from the *Lotu*. It is strictly observed as a day of no work and no play. The guitars and ukuleles are stilled; the radios broadcast only the songs of Christianity; and the day is a day of rest and worship.

This morning I was awakened before dawn by Malé, who had a bad pain in his stomach and asked me to go get his father to come massage him. I walked over to Botoi in the cool predawn light, the village soothed in sleep, and woke Momo; he and Nei came readily, without question. While Momo massaged Malé, I lit the Primus and cut the bread and set out the butter. Nei watched me with kind eyes; she is too shy to attempt even the simplest English; and our Fijian conversation is limited to pleasantries. It was comforting and peaceful in the *bure*, warmed by the familiar light of the kerosene lamp and the spluttering flame of the Primus; and when Momo was finished with the massage, he prayed, his dark strong hands resting on Malé, glistening with coconut oil; and I thought gratefully of this rare relationship of father to son, the healing hands, the healing mind, and of the love they had for one another.

At the end of the prayer, which asked for health for Malé and me and the blessing of the God on the family, he clapped three times in the *cobo*; and we drank our tea and ate the bread and butter, an unaccustomed luxury. Then Momo and Nei went back to Botoi, and Malé slept while I watched the Sunday light softly enter the darkness.

When the sky was light, I ambled over to the outhouse village and sat for a while watching the pigs and the *sulu* flapping in the breeze. On my way back, I passed The Blue Corner—Seine and Qare's old *bure*, where I had listened to Malé play the guitar and where we had spent our first night together. The Blue Corner is where we began our slow journey into intimacy, and I remember that night when I stood outside on the sand, looking at the shadows of the spider gingers and the starry night, feeling the call of the Fijian way.

Now the *bure* is nothing more than half a leaning roof, some posts, and a ragged overgrown border of ginger plants, having been reclaimed by wind and sun. It is a house of fine associations, and it's good to see it slowly returning to air and earth and know that one of these days it will be rebuilt, to catch another net of memories. Often Seine and I joke about the decaying *bure*, "*Isa*, Blue Corner, *isa*!" If Malé, Vili, and Qare ever form a band with matching shirts and hit cassettes, it could be called The Blue Corner Band.

Just next to The Blue Corner is Toa and Sala's eating house; and just next to that is Seine and Qare's eating house, the one we moved into during the hurricane heat; and it too has memories. Seine is up, tending the fire in the corner of the *bure*. Qare, as usual, is sleeping.

The day flaps along like the red curtain in the outhouse. I talk with Seine for a while, mostly joking about the sleeping Qare; then I saunter past our family eating house, where Una is peeling cassava root with little enthusiasm, and I part the bright red curtain with the huge yellow hibiscus flowers that gives some privacy to our *bure*. Malé is sitting on the mat rolling *suki* (Fijian tobacco) into long strips of newspaper, which serve as cigarette papers, and listening to hymns on the radio. We say little—the language is still a barrier between bright talking—and he finishes rolling his smoke and says he is going over to Uncle Saimone's to "ask him something." I know this is a euphemism for drinking *yaqona* and means an absence of some hours. I lie down on the mattress, switch the radio to country Western music taped from the United States, and fall into sleep. I dream of a big tugboat I used to know and am talking with one of the men working on the boat when Malé comes up to ask me how I know the man. It is a dream of jealousy, and I wake up irritated and discomforted, roll myself a smoke, and sit smoking, waiting for the dream to subside. After a while I go off in search of Malé, who is on his way back from Uncle Saimone's, where there "was just a little grog," and ask him to go down the beach with me to the well to bathe.

The old well at Nabau is cool and dreamlike, piled with great dark angular stones and covered over with cut coconut trees that form a platform on which to stand and lift up the water. The well was made long ago by men who knew how to make wells like this, the art long forgotten. The stones are huge, and how they could have been brought to the well is a mystery; they are too big and heavy to have been carried by ordinary men. Malé says the Fijians of the old times were giants. It's hard to dispute him, looking at the stones.

We drop the *tate* (small bucket) into the languid silky water and draw it up, splashing its silvered coolness into the big bucket. Peering into the darkness through the small hole cut in the coconut logs for the bucket, I dimly make out the polished stone walls and the water glinting at the bottom. This well gives life to the village; without it there could be no year-round settlement. It is said that its source is far away on the mainland of Vanua Levu.

Every New Year the people of the village of Lekutu go way up the mountain to the headwaters of the Lekutu River, where there is a great swamp. It is in that swamp that they plant their *dalo*, so each year they dig the *dalo* for the New Year's feasting and plant the tops for the next year's crop. When the water of the well at Nabau turns muddy and brown at New Year's, the people in Galoa know that the Lekutu people have been to Drana, the swamp, to the *vuci* (the wetland *dalo* gardens). A great underground river feeds the well at Nabau; it is said that it never goes dry, no matter how long between rains. In the dry season the water gets low and dirty each day from the constant usage, but each morning it rises fresh and clean again.

When we have filled the big bucket from the well, I take it into a dark grove of cocoa trees set with jewellike climbing vanilla vines. Hanging my towel and clean clothes on a coconut tree, I pull off my sweat-stained shirt and dirty pants and soap myself. Then, dipping the small bucket into the big one, I splash cold water over my body, rinsing off the soap. The sun shines on my body, and I look down at myself, pleasured at being naked, for here in Fiji, if anyone is around, women must stay covered. Finishing my bath, I return the bucket to Malé, who fills it and goes into the cocoa grove. He pours the bucket over his head and sloshes and sprays himself with his usual vigor. My bathing is careful and restrained compared to his abandon.

When we are clean we gather the towels and dirty clothes into a bundle and set back along the sandy path to the village. The path is narrow, flanked on one side with coconuts, and *kou* trees shedding small bright orange blossoms. On the landward side it is dense with climbing vines, spiky *voivoi* leaves, and the big soft floating leaves of the bananas. Great tall coconuts rustle overhead, and we pass under a row of seven immense old mangoes. Then we pass the graveyard in its grove of old frangipani, the fragile sweet-smelling blossoms fallen like stars on the dark ground. The graves are simple rock-bordered rectangles, big ones and small ones, and the graveyard seems untended and neglected. I wonder if it is the Fijian fear of the dead that keeps people away from the graves, for I never see anyone visiting them. Once, just after dark, returning from Nabau with Malé and his sister Una, I teased Una, telling her I would give her five dollars if she would walk on the path next to the graveyard; and she refused, afraid of the *tēvoro*, and took the beach route instead. The only time the graveyard is cleaned is at Christmas, when members of the *mataquali* come with shovels and clear away the fallen coconuts and debris and spread clean sand on the graves.

Soon we are on the path that leads into the village and past the first *bure*. A stand of young sandalwood trees with bright green spear-point leaves looks delicate and appealing. Returning to our *bure*, Malé lies down on the bed and is soon deep in sleep. I walk around the village and see no one. Everyone is asleep.

The day has turned cold, with a strong wind blowing from the southeast, settling slowly into twilight. I go in search of some kerosene to fill our lamps. Finding none, I go into Seine's *bure* to find a bright comforting fire burning under the big black iron teakettle. Seine has set the table, a cloth on the mat with places marked by overturned plates; and we joke about the dinner: fish and cassava. I look into the fire, drawing warmth and peace from it. The fireplace is a simple sand hearth, two concrete blocks holding up two long almost-rusted-through iron bars, with a few pieces of smoke-blackened tin protecting the dry leaves of the *bola*. We talk of putting a fireplace in our *bure*, where I too might draw warmth and pleasure from a fire burning under the kettle on a cold day. Qare moans and turns over in his sleep, and we continue the joke about Qare's sleeping. Then it is time for them to eat, and, declining the offer of *mai'ana* (come and eat), I go back to our *bure* to find that Malé's sister Vaseva has filled the lamp and lit it, and I am welcomed by the light. She does this for us every evening, and I have never had to come into the *bure* without finding the yellow light softly glowing in the darkness. I read for a while next to the lamp, and Malé wakes up and wants to eat. We eat the smoked fish in the *lolo*, and the cassava, and drink some coffee, clear the mat of the two dishes and cups, and go back to bed, snuggling into our separate blankets. I think of Seine and Qare, with only a sheet to cover themselves on this cold night, and of all the people in the *koro* without a blanket or warm jacket to protect them from the cold. We make extravagant love and fall into sleep.

In the middle of the night I go outside to pee in the sand, and see Scorpio glittering above me in a clear cold sky. It seems to fill the night, this great heaven-born fishhook, catching islands and scattering them in the primeval dark.

VAKA VITI

Great white egrets are folded like flower petals on the tiny mangrove clumps that spring out of the shallow waters of Wainunu Bay, a great sheet of protected water that indents the southern edge of Vanua Levu. Wainunu gets its name from a time long, long ago when the first people came over from Viti Levu to explore, and later settle the province of Bua. *Nunu* means to dive down under the water. Here the women in that legendary canoe dived under the water to wash away the ashes with which they had decorated their hair before undertaking the long journey.

The journey had begun in Verata, where Ulu Matua, the eldest son, had incurred the wrath of his father and was forced to flee his village. He and his family were hurriedly packing the canoe for a voyage to an unknown destination when his mother came down the beach, bearing a small plant, protected in a newly woven coconut-frond basket. She tucked the bright green envelope under her son's arm, "Here, take this, and if you reach a place where you can stay, plant it, for the *bula* (thanks for the life)."

Wainunu was not that place, and the canoe continued, reaching Nabouwalu on the eighth night of the journey. The old name for Nabouwalu was Nabogiwalu (the eight nights). They continued down the coast to Bua Bay and headed up the Dama River. There they came to rest in sight of the awesome heights of Seatura, the great volcanic-shield mountain that defines Bua. Seeing *dela ni Seatura* (the top of Seatura), serene and stable above the floating clouds that cloaked its slopes, Ulu Matua decided that this was the place he would stay and found his lineage. Leaving the canoe on the river, he climbed up into the forested heights of Seatura. There he planted the small tree given to him by his mother, in gratitude for the life he had been given by the Gods. The tree was the *bua*, the precious fragrant tree; and the lineage of Ulu Matua became one of the great chiefly lineages of Fiji.

During the times when whales' teeth were impossible to find, a substitute *tabua* was carved out of the dense ivory heartwood of the *bua*, and the *bua* became a sacred tree. *Tabua* has been the supreme sacred symbol of agreement among Fijians since the beginning of their time.

I ask Malé's uncle Navi about the meaning of the *tabua*. His eyes grow misty, focus into the mysterious heartland of the sacred. The *tabua*, the *tabua* is life; it could do anything, anything, kill, keep alive, forgive anything, go against blood kin, bind any agreement. The *tabua* itself, just hanging on the wall, or in a drawer, it's nothing special. But the *tabua* with the *magimagi* cord, given in the ceremony with the right words, and accepted, is the most holy thing of the Fijians. The *tabua* ceremony is very, very sacred, the most sacred.

Uncle Navi tells me that after the settlement of the lands of Bua, a reef was discovered off the point of Naicobocobo which was heaped with the teeth and bones of whales that had gone there to die. The reef was named Cakau ni Tabua (reef of the *tabua*), and that title was given to the chief and all of the inhabitants of Bua.

Our family is related to the chiefly lineage of Bua through Momo's grandfather, and we have come to visit Momo's cousin, the chief of Wainunu. He is going to take us across the bay to pick up a new punt to replace the leaking *Tilaro*. Then Malé and I can move to Vedrala and use the *Dolphin* without the complications of sharing it with the family.

We have hired a lorry to bring back the punt and have driven the long gravel road that snakes through the dryland plains of Lekutu into the cooler wetlands of Nabouwalu and up into the tree-covered mountains of Wainunu. The lorry has gone onto the landing, and we have stopped at the ancient chiefly village. This is the first time that Malé and Vili have come to this village.

Momo will present them to his family in the ceremony of *mata ni gone* (the face of the child) with *tabua* and *yaqona*. Malé's cousin Ratu from Lekutu, a son of the chief of Bua, is also with us.

As we approach the ancient village, Momo bends low at the waist, putting his arms behind his back, and calls the *tama*, the traditional respectful greeting, "*Duo, duo,*" singing the vowels.

From inside the chief's house we hear the long drawn-out answer, "*Oi duo.*"

Around us are the great old *yavu* that once supported immense *bure*. The huge platforms, now empty, recall the time before the coming of the strangers, the Europeans, with their iron knives and muskets and chests of ivory teeth, the time before the coming of the people, and the things, that would change the Fijian way. I feel the presence of that ancient time in this compound; it shimmers in the air, radiates an aura I haven't experienced in any other place. Surrounding the empty *yavu* are large, well-built frame houses, attesting to the wealth of the chiefly families.

It is only in my mind that I can see the great *bure*, the gigantic hand-hewn posts supported by human sacrifices, the great roof rafters elaborately decorated with thousands of yards of red-and-black *magimagi*. Only in my mind do I see the piles of exquisite *tapa*, the gleaming polished clay pots of the cooking fires. These ancient awesome and beautiful things flame in my imagination, then fade into the reality of corrugated iron roofs, planked house siding, and louvered windows.

We file into the house of the chief, speaking softly. The women of the household welcome us and show us into the parlor, a large bright room painted a soft blue, the favorite color of the Fijians. It is decorated with a large net-patterned *tapa*, photographs of family members posing stiffly in studio settings, and a couch, two overstuffed chairs, a coffee table, and mats on the floor. We sit on the mats, first Momo, then Malé and I, Vili, and Ratu, with the women grouped to one side behind us.

The chief enters, sits facing us, is silent, waits. From inside his bag Momo takes out two *tabua*, holds them in his hand, and stretches out the plaited strings, shows them off. He speaks in the strong voice that I am growing used to, a voice that carries the self-assurance of knowing the right way. I listen to Momo speaking, and look at the *tabua* in his hands, at the sweet-faced chief before us, and at the netlike *tapa* that seems to float on the wall behind him. Fortune has brought me into this particular family, where the old traditions still prevail, the ways I was searching for when I first came to Fiji. For *vaka Viti*, the Fijian way, is changing. Most of the villages are compounds of poorly built frame houses. On this journey around the coast of Vanua Levu, from Lekutu to Wainunu, I haven't seen big *bure* like Botoi, and I wonder how often the traditional *tama* is heard in this compound.

The chief receives the *tabua*, murmurs his thanks in a low voice I can barely hear, and leaves. We sit in silence, and another chief, his brother, enters, holding a *tabua*, a very large one, tied with a string of braided *magimagi*. He presents this *tabua* to Momo. "*Vinaka, vinaka*," we all *cobo*. The men recite the traditional answers, the meaning now obscured by time; only the sense, the feeling, remains, "*Ē Ō dua sa dua sa. Ah! Ah! muduō.*"

Later I ask Momo what the words mean, and he answers, "I don't know, I don't know the because, only the way."

This then is tradition, not a questioning of why, but a faithful acceptance of the "is." We say these things; it's enough to say them; we don't have to know the meaning. The meaning is mystery, the meaning is faith in meaning. It is comforting, a relief, not to have to always question the reasons.

When the ceremony is finished, the women bring in plates of

buttered white bread, boiled *dalo*, and cups of tea. The bread is for me, the men eat the *dalo*, and we exchange pleasantries. Momo and the chief pose for a photo, sitting in the two overstuffed chairs flanking the coffee table. The modern furniture seems decoration only, a symbol of wealth, for almost everywhere I go where there is furniture the Fijians prefer to sit cross-legged on the mats.

When we have finished our tea, white bread, and *dalo*, we walk down a steep slippery narrow path to the gleaming river far below us. Malé, Momo, the chief, and I board the chief's big punt; Vili and Ratu will join us at the landing.

We set out down the river into the wide bay of Wainunu. The wind is blowing hard onshore, the tide going out, and the water shallow and choppy. I relax because I am getting used to these excursions into discomfort and realize there is no real danger. I am beginning to appreciate the casual way Fijians approach the details of daily life, a way devoid of excessive concern for safety and comfort, a way that leaves room for laughter and spontaneous joy. So we joke and tease our way along, ducking waves and howling when the spray slaps us in the face. We pass the small mangrove islands frosted with white egrets, dream apparitions in the sparkling sunlight.

After about a half hour we reach the other side of the bay, a forested hill sloping to the sea. There is a rather large compound of well-built wooden houses and a number of sheds protecting identical, newly built boats, all painted blue, differing only in size. Momo was here three weeks ago and picked out one of the boats. It is twenty feet long, flat bottomed, planked with hardwood, and strongly constructed, actually a much more suitable boat for the reefs than our stylish lightly built *Dolphin*.

This compound is the workshop and residence of the Whippy family, a huge Fijian family descended from the original David

Whippy, who came as crew on a whaling boat and stayed to found a dynasty of Fijian-English ancestry. The Whippys are known all over Fiji for their skill in boatbuilding. Members of the family have the biggest boatyard in Suva and are widely respected. I remember Sam Whippy telling me that his nephew Jack, who had helped on my boat so many years ago in Honolulu, had returned to Fiji. He was living in Wainunu. I ask if Jack is around.

"There, he's over there," and I walk over to the sheds, calling.

A large man turns toward me, and I look into the familiar face, clean shaven now and heavier, smiling at me.

"Do you remember me?"

He laughs, "You were the lady with the yacht and the son," and the past floods into the present, imbuing the moment with great meaning.

There is a quality in these events, an atmosphere of rightness, that confirms that I am following a way not to question, a way to accept from the center, from my heart.

SNAPSHOTS

Here is a snapshot of Malé and me taken in Vedrala. We are sitting outside at a low wooden table at midmorning having breakfast. The table is made of old boards salvaged from the discarded punt, painted a soft blue. It is covered with a cloth patterned with yellow and pink flowers. On the table is a disorder of cups filled with coffee, a package of biscuits, a plastic canister filled with sugar, a bottle of coconut oil, a reel with light fishing line, some spoons, a jar of peanut butter, a plate of fried *dalo* root, and scraps of torn newspaper used for rolling Fijian tobacco. Malé is wearing a blue *sulu* patterned with leaves; I am wearing a green *sulu* and a T-shirt imprinted with a picture of

Haena Point, Kauai, Hawaii. The table is on a mat bordered with bright wool fringe. Behind us is our small *bure*. Around us is the cleared and burned ground that will soon be planted with cassava and yams, now a litter of half-burnt coconuts, charred logs, and fallen leaves. The sky is blue-white, overcast, and there is a good wind blowing from the west. Malé is playing with Martin, his favorite puppy, and rolling a Fijian smoke. The other dogs are sleeping: Sonny near the table, Jackie underneath, and Cool in a nearby ditch. The four chickens we brought over from Galoa are comfortably pecking around in the dirt.

The picture was taken in the dream time, and when you look at it you can hear the soft rain sound of wind rustling the coconut fronds and the calling of the *aisau*, a waterlike burble from the bare branches of the *rara*. The *rara* has shed its leaves and is sprouting heavy spikes of silky red-orange flowers. The *aisau* dip their long curved beaks into the flowers and drink honey, mouths outlined with a brilliant band of yellow, pale buff birds with curved yellow feet. Their eyes are bright with the promise of honey waiting in blossoms. Malé says little, concentrating on his smoke and his puppy. His beard is thick, and in the darkness of his face, his eyes shine out, dark lashes contrasting with the white of his eyeballs and the dark brown of his irises. His body is strong and muscled, clean with sun and saltwater and sweat.

I leave the snapshot and walk down the beach to the mangroves. I sit hidden on a curved root, watching the tiny bright yellow claws of sand crabs appearing and disappearing in their hundreds of holes around me. The roots of the mangrove are coated with pale green gelatinous algae and the slow-moving bodies of dark snails. This is my morning toilet, sitting here in the company of miniature crabs and dark-footed snails, in the forest of mangrove root arches. There is always the perfect root, the right height and curve, where I can sit comfortably watching the sun shining on the sea, and leave my used *dalo* and cassava root to mix with the tide.

When I come back, the table is deserted; only the litter of cups and empty plates and jars remain. I clean the table, setting it to to order, and get my typewriter from the *bure*.

A clean page rolls into the platen, and Malé returns, bringing with him a three-foot-long wild yam that he has just dug out in the bush. "I went for the *uvi*, and a bird flew into the ax I was throwing at a tree." He tosses me a small brown huddled thing.

The *aisau* blinks his bright eyes at me and breathes with the quick breath of a frightened bird. I look at his face and stroke his soft feathers and settle him into the folds of my *sulu*. While I describe him on this white page, he rests in my lap, then flies off, a whirr of brown in the dappled green, and is gone to seek another blossom.

Malé asks me what is in the story, and when I read him all of this description, I can see that he is waiting for the action. "Write about the seven crabs we caught last night and how I leave the coconut so we can come and get some more," he advises, mind on food, as it often is.

By the flaring light of the gas lamp, we hunt for the land crabs that live in holes in the sand under the litter of fallen coconuts and dry branches. Our dogs are with us and making a lot of noise, running around in the dry leaves and getting in the way, so it is hard to find the crabs, who, hearing us coming, run back into their holes. When Malé spots a crab, he stalks it, then pins it to the ground with his *sele levu* flat on its arching back, and, picking it up by a back claw, drops the huge ominous thing inside a heavy plastic bag.

Even though the dogs are with us, we get seven big crabs in less than an hour and, on the way back to the *bure*, we bait all of the holes with coconuts that Malé splits open with the *sele*. His raised arm gleams in the yellow light of the gas lamp, the puppies too close for my comfort as the knife cracks into the husk. This way, Malé explains, we can come back tomorrow night and catch all of the crabs that have come out of their holes for the coconuts.

It is all so integrated: the crabs, the coconuts lying right there, the *sele*, and Malé—all part of a perfect technology of simplicity. This is our life—this food, this way of getting it—and we are happy coming home to the *bure* on the beach under the glittering stars, crabs clawing in the plastic bag, puppies stalking sand crabs in the darkness, the gas lamp weaving shadows around us.

Malé brings me the pot with three crabs in it, swimming in the unaccustomed water, bodies cleaned and gleaming in the wet, three crabs of varying shades of pink and lavender and gold, with black hairy legs and alarmed eyes on short eyestalks. "Look," says Malé, "they are drinking water," and I watch the bubbles coming out of the strange crab faces for the last time.

Malé takes the pot over to our kitchen—three stones set in the sand and a one-wall windbreak made of coconut *bola*—and deftly and quickly lights a fire using the *sāsā* and a few sticks. In seconds the fire is crackling bright, and Malé puts the pot on the three stones to boil. When the crabs are cooked, he brings back the pot and sits close to the table. This is the proper way to do it, he explains. If you want to eat for lunch, you must start in the morning; and he cracks the claws with his teeth and carefully picks out the meat, packing it into the body shell. *Mautu*, he explains, is a way work is done with care, not just any way, but the right way, the hard way. He tosses bits of crab claw at the dogs, who, as usual, are lying with their noses two feet away from the food, waiting. When all of the meat has been picked out of the crabs, he will cover the packed shells with sliced onions, thick *lolo* (coconut cream), and squeeze the grated root of the *cago* into the cream. The *cago* (wild turmeric) grows in patches in the shade of the bush, the small orange root the color of carrots, with an exotic and subtle taste of curry. It casts an aura into the packed crab meat.

Here is another snapshot, taken about twelve o'clock, the sun glinting in the coconut fronds, leaves looking wet and shiny in the light. The table has been moved to the shade in front of the *bure*, and you can see the white-fringed hammock from Paraguay hanging between two coconut trees. Malé is squatting near the *vata*. It has taken Malé three hours to pick out the meat from the crabs, stuff it back into the shell, husk and grate the coconuts for the *lolo*, go into the bush to dig the *cago*, and return to grate the *cago* and squeeze its gorgeous orange glow into the pot. Now the crab is boiling briefly, and we will eat.

Our island is feeding us; and, with the exception of the onion, everything—the crab, the coconut, the *cago*, the yams—have come from right here. I think that happiness comes when the material and spiritual details of one's surroundings bring pleasure, and I am really satisfied and happy: with Malé, with the food, with the patterns of color and light that are caught in this snapshot.

WAQA

"Malé," I ask, "do you remember any stories about your grandfather?" We are drinking *yaqona* alone together on Vedrala and waiting for the tide to go down so we can check the fishing net. It is cold. A small fire burns next to the mat laid out under the stars, keeping us and the dogs, who seem almost to be lying in the fire, warm.

If we go to sleep, we will sleep through the low tide and miss the fish, so the *yaqona*, the fire, and our conversation are keeping us awake.

"What about your grandfather? Do you remember him?"

"No. I see the picture, my father tells me the story."

"What story?"

"One Sunday. My grandfather was sick, lying down for months, just lying in the bed in Botoi. He didn't have the pain or anything, just *malumulumu* (weak). Then one Sunday he called all of his children to come: Uncle Tevita, Uncle Malé, Aunty Sereima, Aunty Mili, Aunty Desele, all. They all come, sit in Botoi.

"My grandfather says, 'Today I die. I want you all to take care of each other, love each other, follow the God's way. Poi,' that's my father's name, 'Poi, I give you the land—Vedrala, the compound, Botoi—you work, you take care of your brothers and sisters. Don't fight with each other; fight with the work.' "

Malé looks at me, straight on, says, "That's why my father has the *mana*, because my grandfather gave it to him, gave him the land—Vedrala and the compound, the gardens at Nalomolomo and Namoi—to work, to take care of the family, to love the family. My father, he's different from anybody else. Nobody is like him.

"My grandfather, he was something. He tells the family, 'Don't cry.' Because they crying; Aunt Mili, Aunt Ima, my father, all crying. My grandfather says, 'Poi, I credit one shilling from the *Lotu* (the church); before I die you take the money and pay the credit.'

"So my father, he runs because his father says he's going to die, and takes the money—I think it was fifty pennies—and runs over to the *Lotu* to find the pastor and pay the credit.

"'Then my grandfather, he says, 'Take off my clothes; change my *sulu.*' And Aunt Mili takes off his clothes; they change his *sulu.* She puts on him the white shirt with the long sleeves and the white *sulu va taga* that he wears to church, the nice one, and he lies on the mat in Botoi, and he *lotu* (prays) because he is going to die.

"'My father runs back from paying the credit, and the pastor comes, and the pastor wants to *lotu* for my grandfather before he dies. My grandfather has his eyes closed; he is lying on the mat in his white *sulu*, and he opens his eyes and says, 'Enough, no need; I *lotu* already. Pray finished.'

"'And my aunties and uncles are crying, crying, because my grandfather is dying, and he says, 'No cry, no problem. That's the place we all go; no one can go another way. I just take the lead. The main thing, you take care of one another, love one another, fight the work; don't fight each other. That's the main thing—the work.'

"'They had the coffin, up in the rafters of Botoi. My grandfather had the coffin for a long, long time. My aunty in Taveuni, you never met her, she bought the coffin for someone, but they died before the coffin came, so my grandfather, he ask for the coffin, and they send it from Taveuni. That time, nobody had coffin, just wrap in the mats, or make the box in the village. My grandfather was proud of the coffin. 'That's my house, my canoe (*waqa*); that's where I live when I go in the ground.'

"'Nobody in the family likes the coffin because when they see the coffin, they think of my grandfather dying. But he's not scared, nothing. He jokes about the coffin, 'No worry. That's my house in the ground. It's a good house.'

"'My grandfather, he was something, nobody like him. Nobody knows when they are going to die. But my grandfather, he knew. Sunday. He knew he was going to die on Sunday, and no *rere* (fearful), he takes care of everything: calls the family, pays the credit, changes the *sulu*, lies on the mat, prays. Then closes his eyes. *Mate* (die).

"Nobody else like that. He gave the *mana* to my father, my father like that."

Cool snuggles over to me, puts his head in my lap. Malé reaches over and puts three more sticks on the fire; it flares up, casts shadows on the thick trunk of the *rara*, on the bare branches. "Thank you, Malé, for the story. It's a good story."

Talo (pour a round). Malé mixes the *yaqona*, pours it from high, a murky waterfall from the *bilo* back into the bowl. He ladles out a bowlful and hands me the round polished shell. I cradle it in my palms, look into the opaque mystery of the *yaqona*. The *qona* has *mana*. It's not a drug or a drink; it's something else, just like Malé's grandfather was something else. Just like Momo, just like Malé. Something else.

YOU MISSED YOUR LUCK

Dark, dark in the *bure*. Shadows on the walls like bones. Men's faces, arms, legs, brown skin covering brown bones, flickering in the darkness. Only six of us are sitting around the *tanoa* and the dim kerosene lamp throwing bone shadows on the matting. We are in Uncle Veresa's tiny *bure*. He is telling a story; his voice is low and slow, with long pauses. The words come out one at a time. Uncle Veresa hardly ever talks. Now he is telling a story, slowly, carefully. I understand little—some words, the *tēvoro*, women, the *savusavu*—only the intensity strains through the unfamiliar language; only the bones of the story show in the darkness. Uncle Luke, Momo, Malé, Turaga listen in the long silences of the story.

Momo calls, "*Talo*," and *bilo* after *bilo* is drunk in the angular brown darkness: clap, clap, clap. The story continues. I stare at the orange glow of the lamp, remember reading stories of cannibals. Yet there is no fear here, only the slow intense telling of the story of the *tēvoro* and the women and the sand. I am searching for womanly *mana*, casting in this circle for power to withstand disappointment, power to endure and grow stronger. I sit and stare into the lamp, the fluttering flame, listening.

Later Malé tells me the story. "It's too bad you couldn't understand; it was a really good story," he says, "a really good story. It was when the *tēvoro* came into Uncle Veresa and took his *yalo* (spirit) away.

"Uncle Veresa was young then, about fifteen years old, and he was drinking *yaqona* with his father on Yaqaga. He fell asleep lying next to the *tanoa*; and while he was sleeping, an old Fijian man came up to him and said, 'Come along.'

"Uncle Veresa said, 'I don't want to go with you because you are the *tēvoro* and I am the man, and I'm afraid you might kill me.'

"And the old Fijian man said, 'Don't worry, I want to take you to a really nice place. Come with me. Don't be afraid.'

"So Uncle Veresa's *yalo*, his spirit, went with the old man down the beach at Yaqaga and over the rocks toward the point. Then the old Fijian man said, 'I'm going to leave you here; you wait for another old, old man. Wait here.'

"Uncle Veresa said, 'I'm afraid to wait here.'

" 'Wait,' said the old man. So Uncle Veresa waited. He was really scared, and after a while a really old, old Fijian man came. He was wearing *tapa*. He was dressed in the old Fijian style. Uncle Veresa was really afraid.

"Then the old man said, 'Come with me, and I will take you

to the really, really Fijian place, Na Vanua Yalewa, the island of the women.'

"Uncle Veresa followed the old man along the rocks on Yaqaga to a place which is like the *muni savusavu* (sandspit). He told Uncle Veresa to wait there. Uncle Veresa looked around. The old, old man had disappeared, and he was alone, standing on the sand by the water. And as he looked, the sand changed; it changed into a kind of boat. And the water was going by, going by fast, and the sand was moving in the sea, like a boat, fast. And he went and went in the sea and came to an island. When he came close to the island, the sand disappeared, and he was standing on the beach of that island, an island he had never seen.

"Down on the beach were two women; they were there cleaning the cooking pots in the sea, and they looked at him. They were really surprised to see Uncle Veresa standing there on the beach. They said, 'Come, come over here,' and Uncle Veresa went over to them, but he was really, really scared.

"One of the women said, 'What are you?'

"And Uncle Veresa said, 'I am the *tamata* (human), and you are the *tēvoro*.'

"The women said, 'We have never seen anyone like you before; you are different.'

"Uncle Veresa said, 'I am the man. I have small here (pointing to his chest), and you have big. I have hair on my face, and you don't. I have this, showing his *boci* (penis), and you have this (making a hole and poking it with his finger).'

"The two women said, 'Come with us. We want to show you to our chief.'

"Uncle Veresa went up into the bush with them to the village, and the whole village was only women. The chief was a woman, and the *mata ni vanua* was a woman, and all of the people there were women. And the chief said, 'Come, we eat some lunch.' After they had eaten lunch, the chief said, 'We will go see the *meke*," and the chief and the *mata ni vanua* and Uncle Veresa sat on the mats under the big, big *ulu* tree, and the women came out to do *meke*.

"And the chief said to Uncle Veresa, 'You pick anyone you like for your wife; you can have any one.'

"And Uncle Veresa said, 'I don't want any one because you are the *tēvoro* and I am the *tamata*.' Then the chief called for the *meke* to begin.

"The women came out. They looked really, really nice. They had *salusalu* and flowers in their hair, and they wore the old, old Fijian skirt, the *liku*. And they danced the *meke* and made fancy, winking their eyes at Uncle Veresa and smiling and making this way and that, because they all wanted Uncle Veresa for their husband, because they had never seen a man before.

"But they all looked the same, and Uncle Veresa couldn't pick any one. Then the chief called for the next *meke*, and another group of women came out and did a *meke*. They looked really pretty and made this way and that, but Uncle Veresa couldn't choose one because all of the women looked the same. And it was like that. First the first *meke*, then the second, and the third, the fourth, the fifth, the sixth, the seventh: all came out, and all made fancy because the women wanted Uncle Veresa, but they all looked the same to him, and he couldn't choose any one.

"Then after all seven groups had done the *meke*, the chief said to Uncle Veresa, 'Marry me then.'

"And Uncle Veresa said, 'No, no, thank you very much, but I cannot marry you, because you are the *tēvoro* and I am the *tamata*.'

"So the chief said, 'Okay, you go home then, and tomorrow afternoon late, just when the sun is about to go down, you go fishing by the big stone off Yaqaga. You will see the sharks come around the boat, small sharks, and when you see the smallest shark, you jump into the water and take the shark in your arms and kiss it.'

"Uncle Veresa said, 'Okay, thank you very much. Thank you for the food and for the *meke*. I go now.'

"And Uncle Veresa woke up, and he was lying next to the *tanoa*, and his father asked him where he had gone, because his father could see his *yalo* had gone away. Uncle Veresa was *madua* (ashamed) and didn't say anything. But the next afternoon when everyone was in the *teitei* (garden), he took his fishing line and got into the canoe, and he told his brother that he was going to fish by the big stone on the point of Yaqaga.

"He went in the canoe to the point and was sitting in the canoe when a group of sharks came around the canoe. They swam around the canoe, and he looked for the smallest one, but they were all the same size, and they all looked the same to him. Then another group of sharks came and swam around the canoe, but they were all the same size and looked the same. And another, and another, and another. And six groups of sharks came and swam around the canoe, and Uncle Veresa couldn't pick out one because they all looked the same to him. Then as the last group, the seventh, swam by, Uncle Veresa saw one shark that was the smallest, and he jumped into the water and took the shark in his arms and kissed it.

"The shark turned into a really, really nice woman, and the woman said to Uncle Veresa, 'Come with me. I want to marry you.'

"Uncle Veresa said, 'No, thank you, thank you very much,

but I cannot marry you, because you are the *tēvoro* and I am the *tamata.*'

"And the woman said, 'Don't worry, it's okay, you can marry me anyway.'

"And Uncle Veresa said, 'How can I marry you, because you have no family, and when I take you home to meet my mother and father, they will want to meet your family and know where you come from.'

"The woman said, 'Don't worry about that. I can take care of everything. Just marry me, and everything will be all right and good for you.'

"And Uncle Veresa said, 'No thanks, I cannot marry you, because you are the *tēvoro* and I am the *tamata.*'

"And the woman said, 'Okay, then, it's too bad; you missed your luck.' And the woman touched him on the shoulder and said, 'Too bad, you missed your luck,' and disappeared.

"And Uncle Veresa went home in the canoe to Yaqaga village and told his father the story. And his father told him, 'Too bad, Veresa, you missed your luck. The name of that woman was *Kalougata* (God's blessing).' "

Then Malé looks at me and says, "And my Uncle Veresa really did miss his luck. He married a lot of women, a lot of women, and he never had a baby. And he is married now, and he has no baby. And he borrowed a boat and engine, and the engine broke; and he never has gotten anything. He missed his luck."

And Malé looks at me and says, "It's the old Fijian thing; you don't understand." I challenge the story and ask for details, trying to dismantle it and subject it to reason, to split it with my mind.

And Malé says, "It's the old Fijian thing; you don't understand. It was a really nice story, a really nice story."

⁂ ⁂ ⁂

In the morning, in the netlike sunshadows strained through the woven mat that hangs over the window, I look at the fishscale pattern of light on Malé's arm and think about the story and the *mana* of women.

VEDRALA

Little by little Malé and I are settling into Vedrala. Last week we brought over the pigs, the three brown bush pigs that Uncle Malé gave us and the two white domestic pigs we bought from Toa. Trussed and squealing on the boat, they were sure that the end was near, but when we set them free to make their home on this island, they soon came back to our compound grunting and pushing in search of food and water. The two white domestic pigs are Miss Oink and Miss Oink; the three bush pigs are the bush boy and the bush girls. The male grunts and pushes and shoves the females around, mounting them futilely, for as yet I haven't seen them receive him gracefully. So he grumbles and grunts and heaves, and I watch his sperm drip on the sand. When he is after the Misses Oink, the bush girls chase their rivals away, and it seems that pig life is sleeping, eating, and chasing, or being chased away from favored spots and contemplations.

I try not to think of the Misses Oink or the bush girls coming to an end in the *lovo*. It seems a sad fate for so much fat love and grunting, yet I know that when their time comes, I will submit, just as they must, to the inexorable will of culture. Pigs are eaten mainly at funerals, for not many have the money for weddings or parties, but funerals, if nothing else, must be observed properly. There is no choice but to find, buy, or *kerekere* a pig to feed the mourners at a funeral. Knowing there will be pig or cow meat is often powerful incentive to go to funerals, so they are always well attended: the *oilei* and the screaming of the pigs blending with the rhythmic pounding of the *yaqona*.

Right now there is a blessed stillness in the Vedrala compound. Malé has gone off into the bush to cut firewood to take to Galoa; Vili and Ima's new husband, Robate, are sleeping off a night of *yaqona* drinking; Momo is behind the *bure* weeding; and Ima has gone off to hunt for crabs. There is a soft and comforting easterly breeze; the sea glints and glitters on our doorstep; and the pig population is cruising.

Last week Ima came back with Robate, the boy she had gone off with the night Momo got so angry with us about leaving her and Vaseva behind in Lekutu. Vili never did go and get her; and there were no bad scenes. Robate's family came to Galoa and presented Momo with *yaqona* in a ceremony to make proper the relationship; so all is well. Robate is Ima's second husband; she left her first in Labasa because he beat her. Husband or wife is used to describe the partners of a relationship of domestic, sexual sharing. Some husbands and wives stick together for life, like Momo and Nei; others are more casual and change partners. Ima is casual. She is strong and self-contained and seems ruled more by her head than her heart. As long as the relationship is good for her, fine; when she gets bored with it, she moves on. I like Ima's way. It is straightforward and clean. She is, physically and mentally, a powerful young woman.

Life here on Vedrala is slowly taking its own shape, not the perfect travel-poster focus of slippery lovemaking in the surf, but a more domestic familial shape of pots filled with fish soup and sleeping dogs and burrowing pigs. The past few nights Malé and I slept outside under a big white moon in order to tend the net during the night. The Misses Oink settled down quite near, the dogs competing for the pillows, and a bright fire kept the teapot ready. Momo and the rest of the family have been coming over during the day to work on the island, leaving us alone at night, a blessed compromise between Malé's need for family and my need for privacy.

I look at the Oinks with longing, would love to nestle in next to their bristly bulk and bed with them. They are still too wild for that association, and will hustle off if approached too closely. Pigs, when left to themselves, are clean, and ours take three or four baths daily, sitting in the shallows, big ears swinging lazily, truly contented. The animals seem to be finding their places with one another: the pigs and dogs and chickens observing the proper forms, fighting a little for food or water, but basically companionable.

A few weeks ago Malé built me a small *vata*, a shelf made by lashing mangrove sticks together on a framework of posts set in the sand next to the cooking fire. It makes cooking much easier, holding the pots and cutting board (a piece of driftwood lumber salvaged from the tide) and giving me a place to keep the cooked food out of reach of the ever-present pigs. Our kitchen started as three rocks laid in the sand to form a tripod support for the kettles, and then a windbreak of *bola* mats lashed to sticks set in the sand was added. Yesterday Momo made a lean-to roof over the three stones and added a windward wall. Now there is a semblance of a kitchen—fireplace, table, roof, and one wall—all serviceable and kindly to the eye.

Just past the kitchen a fire has been built on the high-tide line, topped with a wire supported on iron bars pushed into the sand, and the fish are smoking for this coming Sunday's dinner.

Last night Malé and I went out to tend the net under a brilliant warm sky, the net more holes than mesh. Momo says it's luck whether the fish go into the mesh or through the holes. Even in its terrible condition, with patches big enough to permit the passage of porpoises, the net manages to catch thirty to forty pounds of fish of good variety. So we have enough fresh fish, and the dogs and pigs can eat fish soup and boiled fish. Nothing is thrown away except the scales; the guts are cleaned and boiled along with the fish for the dogs and pigs.

Tending the net on a warm moonlit night is softly seductive. Malé poles the new punt through the shallows, and the gas lamp turns the water in front of the boat a milky green, mysterious and beckoning. Last night we saw an immense stingray next to the net, and Malé threw his spear at it. Then we poled after the dipping, weaving, dancing spear. The rays are killed because they can really hurt; the flashing tail could open your flesh to the bone if you stepped on it. Chasing the big ray in the darkness was a little scary, with Malé balanced on the bow, lunging after the gliding spear. After about six or seven tries with the light spear, he gave up the chase, and we returned to the quieter pleasures of taking fish out of the net. The ray was as wide as my two arms held out sideways, the tail about four feet long, and lethal, a great dark spotted butterfly of the shallows.

It's absolutely quiet off Vedrala in the night; ours is the only light. There is no one else on the island but us and the dogs, pigs, and chickens. The land hums and sings in the pale light. It is only when it is really, really quiet that you can hear the land itself singing. I love it; and Malé is a bit scared, for he has been raised with the *tēvoro* and fears the uninhabited night. I see that for all of his strength and courage he is soft inside.

I am beginning to perceive that beneath all of the confusion there is a balance between us, something proper and useful to both of us. Yesterday Malé told me that he was glad he had married me, that if he had married a Fijian, she would be angry with him all of the time. He would always be ordering her around—telling her to go do this and bring that and get this—and I won't let him do that to me. He is beginning to see an advantage in gentleness; and as his temper and style soften, he helps me grow stronger physically and psychically. He is teaching me ways to do things more easily: start the fire, cut the firewood properly, tie it so it can be easily carried, strip the bark from the *vau* to make instant string, catch fish, pole the boat in the shallows. Most important, he is teaching me to *vosota* (endure), stop whining and thinking how weak I am and make my mind strong so that my body becomes willing.

Now little Malé strolls by, completely naked, his small strong body the color of the shadows. He is followed by small Suli, and they are having an adult conversation about who is going where, which seems to be the main domestic conversation in the *koro*. Suli climbs a few feet up a leaning coconut, decides it's too hard, and runs off in a very important manner. Malé follows, looks at the coconut, decides that it isn't as interesting as Suli, and runs off after her.

A pig sniffs at Robate's feet, and Robate wakes, stretches, picks up his *sele levu*, and heads into the bush to cut grass to fix Vili's *bure* roof before the rains come.

Small Sione appears carrying a long heavy branch, followed by big Malé carrying two heavy logs. Malé drops the two big logs, the size of small trees, with a heavy hardwood clunk, onto the growing pile of wood, lying there to dry and season. He is planning ahead, knows that when Christmas comes he will want to drink *yaqona* all day and night for three weeks or so and is taking care of the wood supply now so he won't have to work later. He stands, his back to me, looking at the blue glimmering sea, the big silver-barked tree limbs at his feet, dark shadows of the leaves in the foreground.

Clink, clink, Momo's *sele* bites the thick grass. The *aisau* call, the *rara* has shed almost all of its brilliant orange blossoms and stands bare branched, waiting for the rain.

Malé comes up to me with small Malé, says to small Malé, "Show Joana the *cina* (lamp)." He pulls back the skin on small Malé's boci, and his little penis pops out.

Small Malé laughs and slides his eyes at me and says. "*Cina*."

Malé explains that small Malé called to him. "*Ta* Malé, want to see the *cina*?" then pulled back his foreskin so that his little *boci* popped out, laughed, and said, "*Cina*." So, of course, Malé had to bring him here to show me the trick—something small Malé had thought up by himself.

Small Malé is a fascinating charmer of endless seductions. When he wants to eat something special, he rolls his eyes at the food and then at me. He can make one eye roll up and the other sideways. His eyes are huge dark irises and big round whites; and he uses them like a master. When he is angry, he will stare at big Malé forever, a fierce look of pure hatred anchored in his dark eyes. When he wants something, his eyes become hands reaching out for the delights of peanut butter or chocolate cookies or fried peas. We call him the "*Bula keke* boy" because whenever we return from a trip to the mainland with possible treats, he rushes down the beach to greet us, calling, "*Bula*, Joana, *bula*," thinking only of creme-sandwich cookies, which he calls "*keke*."

DRI

The big story on Galoa these days is the *dri*: the great *dri* epic. The *dri* are sea cucumbers, sometimes called *bêche-de-mer*, and during the last century there was considerable trade in Fiji in *dri*. The Yankee masters would contract to deliver *dri* to the Orient. There the dried creatures were soaked in water, cooked, and became the essential ingredient in a rich soup, much favored, so I read, by the Chinese. During the days of early foreign contact, before the suppression of tribal wars, the *dri* trade was perilous. Often the locals, resenting interference in their traditional way of life, revolted, burned down the houses that were built to dry the *dri*, and disappeared into the bush, leaving the Yankees *dri*-less. The houses were hot, smoky places; the work was boring and demanding, and was ordered by the chiefs, who received cheap muskets and trade goods in exchange for cargoes of *dri* worth thousands of dollars to the Yankee captains. The men and women ordered to scour the reefs for the *dri* and prepare them received nothing but watery eyes and aching backs for their labor. Soon the *dri* were gone from the reefs along the coast of Vanua Levu, and the Yankees departed, to return a few years later to strip the coast of its precious stands of sandalwood trees.

Now the *dri* are back, dotting the reefs like so many dark shadows in the corals, and the trade has been reopened through markets in Suva. Almost everyone in the *koro*, tantalized by fabulous stories of hundreds of dollars being paid for bags of dried *dri*, is out combing the reefs for the big black sluglike creatures.

There are very few ways for village Fijians to make money, and money is needed for store-bought essentials. Fishing sometimes brings a little extra cash, but sometimes it doesn't. Making *copra* requires too much labor for the sixty dollars or so one might get for a 100-kilo bag of dried coconut. The making of *dri* is one way island families can earn a few extra dollars to buy necessities and perhaps replace a worn-out gas lamp or buy a blanket or a new *sulu*.

At night the beach that fronts the *koro* is dotted with glowing fires that from a distance look romantic. Momo has constructed a *dri*-drying enterprise on the beach from a 44-gallon drum, designated for our improved outhouse, but borrowed for a more important venture. The drum has been cut in half lengthwise and is propped up over a roaring fire. Hundreds of scummy *dri* boil inside. Malé brings me a mat; we bring out the guitar and ukulele and pass the time waiting for the *dri* to boil. Nei, Ima, Una, Banoko, and Suli sit on an old mat cutting the *dri* in half and scraping out their slimy insides. It is tedious and smelly work. Vili stirs the boiling *dri* with a long stick. When they are completely cooked, he dumps the whole drumful out on a bed of coconut leaves, where they cool enough to allow Nei and the girls to cut and clean them. Then the split, cleaned *dri* are laid out on a wire built over a low smoky fire, where they will undergo the first drying. The making of *dri* is truly a family enterprise: gathering the wood and tending the fires, picking the *dri* during low tide, boiling and drying the *dri* the same night they are gathered, and then the next day the sun drying them on racks along the beach. It takes a whole day and night and then part of the next day to sufficiently dry the *dri*. The finished product looks like a small flat lump of black leather and is stored in empty rice bags.

We joke about the *dri* and the propensity of the *koro* to chase rainbows in the form of promised riches. The story is that the *dri* will bring $2.40 a kilo, or about $200 a bag. Fishing is neglected; the Peace Corps fish market that Ben labored over is in limbo because no one is fishing, preferring to make *dri* instead; and the precious scanty firewood of the island is depleted to boil the *dri*. I wonder aloud what will happen when the *dri* and the firewood are gone. How will people cook their food?

In their quest for the satisfactions of the moment, Fijians often neglect the future. The firewood is a case in point. No one seems to be interested in planting trees to replenish the firewood supplies, and Ima, Una, and Malé have to go farther and farther into the bush to gather, stick by stick, the firewood for cooking.

Thankfully we have Vedrala and can take the boat over to the island and gather wood for cooking on Galoa. But other families are not so lucky and soon may have to face the serious fact of very little firewood. By then whatever money they will have earned from the making of the *dri* will be long gone; the fish market will have collapsed from lack of attention; and there will be no place to market the fish except for Munci's store. Mr. Munci is a serious businessman, and once he is the sole purchaser of fish, he can control the price. But the making of *dri* continues, and so does the speculation about the price to be paid for the horde of *dri* waiting conversion to gold.

Today the first *dri* returns are in. The bags of *dri* that according to last week's story were going to bring $2,800 to Saraqia's family, brought $400. When Saraqia took the *dri* to Labasa, the price was down, so they brought the *dri* back and decided to take it to Suva. Reaching Nabouwalu to load the bags on the boat, they found out that there was a surplus of *dri* in Suva, so they returned to Galoa. Then they heard that there was another agent in Labasa who paid a big price, so they took the *dri* back to Labasa. Looking at the bags, the agent promised them $2,800. The story swept the *koro*, and last night Malé assured me that they had the money. Then this morning Momo brought the latest news. The agent received only $400 from the buyer in Suva, and the $400 was divided among all of the various people in Saragia's family who had helped make the *dri*. Trying to find out the who, what, why of the story, I find out nothing. No one really knows how many bags or who took the *dri*, but everybody did know that they were going to get very, very rich on *dri*. Now, looking out on the reef, I see the punts from Tavea collecting *dri*. I guess they haven't heard the story yet, or have one of their own.

Not long ago a rather impressive sport-fishing boat came to anchor off the *muni savusavu*, and three men came to the village:

an American and two Fijian helpers. The village men gathered in the chief's *vatu nu loa* to listen to the story, lured there as much by the *yaqona sevusevu* as by the strangers. The American explained, through his interpreter, over the grog bowl, that he was there to develop a market for lobster. His intention was to buy the lobster from the villages of Galoa, Tavea, and Yaqaga and air ship them to markets in Hawaii and the United States, where they would bring high prices. What was he prepared to offer for the lobster, the men wanted to know. Maybe as much as three or four dollars a kilo. It sounded really good, and the *yaqona* tasted sweet, so it was an easy matter to promise to supply the American with hundreds of kilos of lobster a week. Sure, they could get lobster, knew all of the lobster grounds; it was no problem. But there were some things they needed: iceboxes, for one, to store the lobster until the American or his helpers could pick it up, and at least a drum of benzine for the boats, so they wouldn't have to waste money traveling to Nakadrudru and paying Munci's high price for benzine. It all sounded wonderful, but I viewed the conversation with some reservations, having never seen more than one or two lobsters a week in the village.

True to his promise, the American supplied the iceboxes, the ice, and two 44-gallon drums of benzine for the men to use fishing for lobster. The cost of the benzine would be subtracted from the price paid for the lobsters. To help out some more, the American supplied some fishing spears, which were to be signed for and again paid for out of the lobster proceeds. A few weeks went by.

There was a funeral in the village that occupied everyone's time: gathering the food, going out for *vonu* and fish, preparing and consuming the *magiti*, and drinking *yaqona* for three days and nights in a round of ceremonies. No one much bothered about the lobster. When the American came back to pick up his lobsters, he found the benzine drums empty—they had been *kerekere*'d for the fishing—and two small rotten lobsters in the bottom of an empty icebox. The spears had disappeared. There were no names on the list.

It's not deception. It is just that the priorities of the two cultures are vastly different. It's easy to promise lobster around the *tanoa*, and the manners of Fijians demand that they say, "Yes, of course," even though no one really believes in the yes. Making money is important, but not as important as fulfilling ceremonial obligations. And the spears? Well, what does it matter; it's a good joke.

PIGS

The pigs are becoming completely tame. They follow Malé into the bush and along the beach with the dogs, an absurd procession: Malé in the lead, carrying his *sele levu*, calling oink, oink, oink, followed by five pigs and our five brown dogs chasing and playing. Malé's cherished chore in the mornings is to go into the bush at dawn with the pigs following and cut open coconuts for them to feed on. If there are no fallen coconuts and he doesn't feel like climbing the trees, he simply hacks off a strong stick from a nearby tree and hurls it with astonishing force up into the coconut tree, dropping the coconuts. I marvel at his strength and skill. It is not easy to hurl a stick up into a tree with enough strength and accuracy to break the thick stem of the nuts and send them cascading to the ground. The pigs are wise to these early-morning feeding expeditions. As soon as Malé picks up his *sele*, they grunt to their feet and follow him for breakfast. Yesterday afternoon Malé went hunting for fish along the reef with his spear and was followed way out into the shallows by the dogs and pigs, a Fijian pied piper of pigs.

The pigs sleep by the fading embers of the fire, nestled

against the windbreak *bola* of the kitchen lean-to, fat slack body next to fat slack body, big radar ears motionless, restless noses stilled. The Misses Oink are always together. I never see them more than five feet apart. Pregnant and heavy now, they are hardly recognizable as the sleek pink piglets that used to burrow next to the outhouse on Galoa. The two bush sisters keep to themselves. The bush male cruises, sometimes alone, sometimes with his bush sisters, and sometimes with the two Oinks. Malé and I joke about the Fijian pigs and the European pigs. Malé has a decided preference for the white Misses Oink and will protect them when the bush girls get hostile and try to chase them away. All of the females are in heat, or season, or whatever it is that pigs get into; and the male mounts them, or tries to mount them, often. Malé and I watch unexpurgated pig sex with open interest and satisfaction, tossing in words of encouragement in an endless sexual teasing that centers on who gets what satisfaction. The pig's penis is thin, like a pencil, and over a foot long and bright red. It emerges like a corkscrew and searches for its target, seeming to need no assistance to poke into the tiny opening barely offered by the female. When inside, the male has that blissful look assumed by males when they are sexually content. His eyes close halfway, and his face is slack and almost smiling. He holds the female tightly with his forelegs and rests all of his considerable weight on her. The female slides around for a while under his weight then is still, her eyes half closed; and the two remain that way for a number of minutes—silent, unmoving. Only a little twitching from the male's haunches and a little grunting from the female show that they are conscious. There is none of the convulsive jumping around and moaning that has come to signify sexual satisfaction, and I wonder if the female likes it. She must like it, because when she doesn't want the male, she screams and bites him and turns away, and he cannot enter her. All of this speculation about pig sex turns us on, and we go off to the *bure* to try some of our own.

I wish my mind were as clear as this Miss Oink's is, for mine is constantly finding fault with and looking for holes in the fabric of perfection that I would call love. Oh! just to stand still and half close my eyes and, when finished, lie in the shallows cooling my belly or burrow in the cool dirt under the trees, body and soul at rest.

The vocabulary of the pigs consists of low grunts and high squeals, which often sound angry. The grunting is almost continuous, a purring, which then will change into a screeching squeal of fury or demand. The pigs sound very much like the whales. This is the first time in my life I have been around pigs, and I find them charming, but potentially fearsome. They are big and strong and determined. The male is beginning to grow his tusks, and when he and they are full grown, he will be formidable and dangerous. Right now the pigs are winsome, but I get nervous when I think of increasing our herd and imagine twenty full-grown pigs rooting and grunting and tugging at the water buckets or overturning all of the pots and plates on the *vata*. Both Malé and I love raising animals, but neither of us is very eager to kill them. We will have to fence part of the island for the pigs, or fence the gardens, either of which will be a big job.

According to the map in the Native Lands Trust Office in Labasa, Vedrala is about forty acres. There is small hill attached to Vedrala by a saddle of mangroves, which is another five acres, so the island, with its surrounding mangroves, looks much bigger when seen from the sea. Momo plans to clear a considerable portion of the island for gardens and ultimately to build us a big traditional *bure* so that we can settle here permanently. He and the rest of the family will continue to live in Galoa, but if he has the energy he will build a small family *bure* on another part of the island for their use when they come here.

Fijian land use and tradition are perplexing to the European mind. It is ownership, but not ownership. Our *mataquali* owns Vedrala, and as such the island is available for anyone in the *mataquali* to use. By tradition whoever first clears *mataquali* land has the right to use it. No other member of the family will come and try to plant cassava in someone else's garden or build a *bure* in someone else's compound. Thus, even though Momo's father has given him Vedrala, Momo must follow Fijian custom in determining its use.

Clearing the island is heavy labor. The head-high grass and small trees have to be cut down with the *sele*. The grass is left on the ground to dry, then burned. The big trees and coconuts do not burn, but the brush and grass do; and after the fires have died, the ground is raked and cleaned of charred coconuts, logs, and rubbish. Since Vedrala has been uninhabited, the soil is soft and fertile, unlike the garden plots on Galoa that have been intensively cultivated.

I look at the cleared ground extending around me and realize that slowly, over these last months, the land is being tamed and made useful. I remember that first day when Malé and I came to the island in the rain, the bush an undecipherable puzzle, the future blank and confusing. Now our life here is becoming a reality.

When we have cleared, fenced, and planted our gardens; established our pigs, chickens, and dogs in mutual harmony; and built our big *bure*, we will be basically self-sufficient. We have coconuts in abundance, red-chile-pepper bushes, yams, turmeric, a rich reef to fish, and a family with the skill and desire to care for one another. When I met Malé, I had no idea that he and his family had their own island; now I realize that I am particularly fortunate to have met him.

Malé has taken the *Dolphin* over to the village to bring back water, get cassava from the garden, buy cigarettes, and bring back the ax. I am here with the pigs and dogs and chickens. I want to sleep in the hammock and listen to the rustle of my mind and the soft sweet sound of high water lapping on the shore. It is cool and quiet, time for a little travel-poster laziness.

It is almost dark when I hear the high whine of our outboard announcing that Malé is returning from the *koro*. I walk down the beach to the point to greet him, our dogs my instant companions. Cool is aptly named, for he is the cool one, the self-possessed hunter, always fascinated by the world around him, way ahead, chasing sand crabs or splashing in the shallows. Martin is possessed only by his desire for food and sleep. He is the lazy, competitive one. Jackie is timid and gentle, the first to appear at my side and nestle close to me. I am amazed that these three dogs from the same litter, raised in the same way, can have such different personalities. Sonny, their father, is hard-headed and contained, yet eager for affection. We wait at the point, watching the small white speck far offshore grow bigger and noisier, the dogs racing along the sand to greet Malé as the boat comes into shore.

* * *

This is the cherished time: just at sunset, when we work together in the cool of the evening to gather and burn the dry leaves, charred coconuts, and branches left from the initial clearing of the bush. Malé strikes a match, holds it in his cupped hands to nourish the flame, touches it to the spray of dry coconut leaves at the end of the frond. The leaves glow; then a tiny flame runs out and catches on the crisp spikes of the *sāsā*. Malé bends the tip to the ground, the flame runs up the leaves, flares. He touches the flaming branch to the huge pile of dried leaves. The pile heaps into light, an intense heat rises, drives me away from the bonfire, back onto the cool sand. Behind me Yaqaga is a dark sharp mass against a flaming sunset sky. The colors are the same: the fire on the island, the sky blazing with red light. It is silent. We do not talk, only search in the growing darkness for more debris to pile on the fire. Soon we have a string of great fires blazing behind the coconuts that line the beach: a roaring splendor of firelight in the dark sky.

Somewhere out there in the black night, Ima and Momo pause in the mangrove shallows of Vatoa, ready to lay the net, the gas lamp casting gold shadows on the calm sea. Momo poles from the bow; Ima carefully feeds the net out of the side of the boat, letting it slide into the quiet water. It is high tide. Momo looks toward Vedrala; the shoreline is a string of flaming points. The fires look big; they warm him. Momo feels good; he looks at our island and sees it alight.

VATOA REEF

"**Look!**" Malé shines the flashlight on the dark, stiff leaves of the mangrove, stained and spotted with white droppings from birds. This is a rookery, a staying place for thousands of sea terns who shelter in Vatoa reef during the night, then leave, in vast long looping spirals to fish on the deep sea during the day-time. Malé shines the flashlight on a sleeping tern, inches away. The bird is stupefied, blinks into the light.

Malé scoops the bird from the branch and hands it to me. "Here." A warm quivering in my hand, a hint of dark gray banded with white, confusion in its eyes at this unruly awakening. I slack my hand; the tern flies up and away to find a calmer place to continue sleeping.

At night the terns shelter here, small feet clawed around the thin branches, perched above the shallows. During the day, it's the fruit bats, thousands of them, motionless, folded like velvet cocoons deep in the mangroves of the reef. Bats by day and terns by night, changing places. The terns sleep on the margins of the reef, the bats deep inside. At dusk the bats fly out to the islands—Galoa, Vedrala, Tavea, Yaqaga, the mainland—to settle in the breadfruit, the papayas, the mangoes, and feast on the pulp and juice. Fruit bats, clean; their flesh tastes of fruit—sweet and juicy. I have not yet brought myself to eat one, having been raised on comic book and movie bats, but Malé and Vili and Qare hunt them eagerly in the *koro*, even leaving the *yaqona* circle to hurl sticks into the *ulu* trees in the compound at the feasting bats.

I am appalled at these midnight hunts, and think them cruel, but I am assured that they are wonderful eating. One night Malé brought me a wounded bat, hung it on the top of the mosquito net, a poor crippled thing with one broken wing, huddled there, trying to escape our curiosity. I begged him to kill it; there was no way it could live, but in the morning Malé hung it from the *baka* tree, like a sad dry leaf, and soon it died, still resolutely hanging by its one good wing.

There is a magic in the flights of the bats as they float into the islands at dusk looking like big birds, silently settling in the trees. There is something primeval in their hushed flight, something that touches the senses, enchanting.

Malé says the terns go far out over the deep sea, where they fish all day, returning each night to settle in the reef. He says he has seen the water black with resting terns bobbing in the swells, from a distance a great dark mass resolving itself into hundreds of birds. Before this time, when the Polynesians and Melanesians still remembered the vast intricate lore of deep-sea navigation, the terns were a sure sign of nearby islands, an infallible natural loran. Now no one here travels in that ancient way, and the birds are a charming curiosity, or at best the sign of schools of bait fish, signaling the presence of bigger fish, such as tuna or king mackerel.

But for me now the terns are an opening into love for Malé, the nonintellectual naturalist, my dark prince; and it is when he is doing something like stalking a sleeping bird, to catch and show me, that my love flares and lights up the night.

Tonight it is dark, very dark. A small moon will rise late, but for now only the flashlight and the yellow glow of the Coleman lamp show our way. We have come to set the net in the leeward shallows of Vatoa reef. The wind is blowing, but here on the far side of the reef it is calm and warm. The net is piled in the back of the *Dolphin*, and I play it out while Malé poles the boat along the edge of the mangroves. It is not a smooth operation, for my technique leaves much to be desired, and Malé is not shy about telling me about my shortcomings as a fisherwoman. Finally, the net is laid, the lead sitting on the bottom, *vau* floaters bobbing on the surface; and we return along the two thousand yards of the net, checking to make sure that there are no snags or places where the fish can swim out.

Then it is time to wait. There is little to do but anchor the boat just outside of the net and wait for the tide to go down and trap the fish. We have no Primus stove and no tea or food to cook, so there is little to occupy us during these long, slow hours of waiting for the tide to go down. Besides, our squabbling over my technique of laying the net has left a strain, so I curl up on the floor of the boat, pillow my head on my arms, and cover myself with a jacket, while Malé rolls and smokes Fijian tobacco to keep himself awake.

The terns occupy my mind: soft gray fluffs of intensity staring into the pinpoint light of the flashlight. They are not far away, sleeping in their sanctuary of leaf and tide. Vatoa is a marvelous landscape, a great twisting shallows of mangrove clumps and channels, opening like flowers into transparent pools, residences of astonishing creatures, defying description. Creations of flesh informing the shallows. Coral is a dead word, conjures images of bleached white bonelike things sitting on coffee tables or on glass shelves. Here there are corals that look like lips and like the inside of bodies—fleshy, undulating, spongy to my touch. Hiding inside the great flat mushrooms of coral heads are the *vasua*. They open frilled and seductive fleshy lips to the currents—slack, subtly striped, royal purple, soft brown and green and beige invitations. Once you have recognized one in its camouflage, the rest become easy to spot. They are secrets of nature, painted by the hand that fashions this canvas of the alarmingly beautiful forms of possible life. Tiny electric blue fish guard and dart through this landscape of flesh that, slowly sucking and exhaling, causes almost imperceptible currents and eddies. Shells imitate the colors of low-tide mud, grow shaggy coats of dull brown grass to further disappear. Yet their flung pointers reveal them immediately in front of my foot, when I stop looking for them.

The tide settles, goes somewhere else, swells the ocean in a distant place, and the *Dolphin* floats down into shallower water. There is a half bottle of wine on the boat; we share it and speak very little. The night is vast; the darkness expands as we sit watching the stars, silent, somewhat cold, in the presence of the sleeping terns. The tide flows away from the reef; the net settles, like a fence, low in the water; the *volaca* flatten themselves in the shallows, go among the forest of roots, waiting for the water again. It is time to check the net.

The water feels cold as I slide into it from the bow, not much higher than my thighs, but cold. I watch the pool of light in front of me, hold out the gas lamp to see, and pick my way carefully through the sleeping corals, their daytime splendors diminished by this light. Malé is ahead in the darkness, surefooted and quick. I stumble behind, cautious, off balance. "There, there!" shouts Malé. "Go there!" gesturing in the darkness toward an unseen destination.

I stumble vaguely in that direction, holding up the gas lamp, unsure of why the sudden hurry. "That way, that way!" shouting again and running. I don't know what "that" way is, seeing only the circle of light at my feet, and splash around uncertainly in some direction toward the mangroves. Malé runs along the net, unties one length and pulls it in toward the mangroves, making a short, curved fence. Somewhere there is a school of fish, a big school, and he is trying to herd them into the net. By the time I understand what he is doing, the fish are gone, have run back into the mangroves to wait. We slosh back to the boat.

A thin flat glow of red light pervades the dark above the mainland. The air is really cold. It is the time before the dawn when everything seems still. The late moon is rising, golden flaming orange in the dark sky, one brilliant star above her. She assumes fantastic shapes, makes her way above the mountain, slides through the clouds, floats free into the sky. Dark hushed shapes of birds come from their sleeping places, turn toward the sea, silently. The tide is low, it is almost dawn. The sky reddens beneath the rising moon, the star high and diamondlike in the bluing sky.

We slide back into the cold water, Malé ahead of me, impatient to get at the fish. There is just light enough to make out the shadows between the corals, for Malé's practiced eye to decipher the shapes of hiding fish. He cuts a mangrove root, peels and splits it in two, ties the two strings together, and tosses it to me. It is time to work.

The fish are everywhere, hundreds of them, waiting out the turn of the tide—trapped behind the net, hiding in the mangroves, flattening themselves in the sand to disappear except for a faint disturbance on the bottom. Malé has the spear, and I have the string. We start.

"Here," he tosses the first fish toward me. He continues, spearing them out of their hopeful somnolence, poking them through the eye to kill them. "Here," he tosses me fish in the warming light: big fat flat fish that slowly float to the bottom, settle on the corals. I run after him, picking up the fish, stringing them through the gill slit and out through their open mouths, carefully, for they have back and fin barbs: and, barely dead, their poison is strong enough to hurt for hours. Malé throws the *moto* (spear), running through the shallow water, little silver splashes following him. The string grows heavy, is soon full, and he pauses to cut another. I leave the strings of fish lying in the water; we will pick them up later when we are through. Now there is no time, for the tide has turned and will soon flood in, freeing the fish, allowing them to go farther into the reef, until the full tide liberates them into the open sea. We are working fast, yelling and laughing, linked together by our common perceptions—spearing, tossing, stringing, no time to waste.

Full daylight comes, flattening the golden colors of dawn. The night's magic has fled into the open colors of daylight. But we are catching fish, lots of fish, and we are invigorated by our action, by our success.

An hour and a half passes with no rest from the spearing and stringing. Fifteen strings lie scattered around us. The water is too deep to continue; the *volaca* have gone into the reef; it is time to leave. I collect strings, dragging them behind me, still careful not to get brushed by a barbed fin, and pull them toward the boat. Malé piles up the strings inside the rising net near the *Dolphin*, secures them to the *moto*, which marks their location. We head down along the net in hip-deep water to retrieve whatever has been caught in its meshes; big mullet float like fat silver cigars, aimless, stuck. They are easy to disentangle, have no spines to worry about; just unhook the mesh from around the gills and slide them out.

I pull the *Dolphin* alongside of us, holding onto the gunnel while Malé frees the fish, tosses them inside. Had I not been so uncertain last night, we might have captured the school, driven the two or three hundred mullet into the net. As it is, we have snagged a good fifty, and the silver pile gleams. We wade back to the strings of *volaca*—fresh and heavy—load them into the *Dolphin*, cover the pile of fish with wet T-shirts and towels, and get going.

The sun is higher now, and it is getting hot. We have to hurry home to clean the fish and get them to market before they spoil. We leave a frothy plume behind us and head toward Galoa. Somewhere inside the reef the bats are motionless and silent, full of sweet juice and pulp. Somewhere far outside the barrier reef the terns are fishing.

SEED YAMS AND PIG CHASE

Today Pipi set a fire to burn the grass off the margins of his plantation and in so doing cooked all of Toa's seed yams and *uvi*, which were stored at the edge of Toa's adjoining plantation. Toa will have no yams or sweet potatoes to plant this month, and so will have to eat only cassava and *ulu* all next year.

We are sitting around the *tanoa* in front of our *bure*, where Malé, Uncle Veresa, Cousin Luke, Vili, Koroi, and Rusi are discussing the day's events. Toa keeps shaking his head. Pipi is tall, exceedingly ugly, and not really smart. There is nothing much Toa can do except go to Pipi and ask him for some seed yams.

After the yams are harvested, certain choice ones are piled together and covered with grass, or stored in a small *bola* hut erected for them in the garden, and kept over until the beginning of the wet season, when they can be planted. These yam and *uvi* stores are precious, carefully watched; and it is serious that Pipi has destroyed Toa's yams. Luke is making a joke of it, telling me that now Toa doesn't have to cook his yams, only go to the garden with a basket and get the already-cooked yams. Toa shakes his head.

The big story today is the pig. For weeks and weeks the men have been going into the bush on what seems to me a comical hunt to capture some village pigs that have gone wild and are uprooting the cassava and eating the seed yams. These hunts are totally disorganized, accompanied by a maximum of noise and laughter and chasing, and never seem to result in getting the pigs. The hunts are always well attended, for they are an interesting alternative to real work.

Today, leaving the outhouse on the *muni savusavu*, I heard the hysterical pitch of women's voices and looked up to see Sala, Seine, Marselina, and several other women standing on the shoreline at low tide screaming. The men had chased a pig into the water and were running in the shallows brandishing spears and shouting. Then I saw our boat just outside the reef and ran toward the boat, screaming back at the women that the way to take care of pigs was to be kind, to tame them, not run after them with spears. I floundered through the coral yelling at Malé to wait for me, because he was apparently taking the boat off somewhere when he had just promised that we would spend the afternoon alone together in the bush, cleaning the banana plantation.

I reached the boat to find Malé, Momo, and Una all excited and ready to chase the pig, who had fled from his tormentors into the deep water. I told Malé that I wanted him to stop, put down the anchor, and have a talk.

"But, Joana, the pig is swimming away!" and Malé powered out in search of the pig.

It was just like a turtle hunt, but I was in no mood for hunting anything, much less a harassed pig. Momo stood poised on the bow, turtle spear in hand, and Malé turned to me, ecstatic, "Look, Joana, it's just like the *vonu*."

"Malé, no! Let the pig go. Don't kill it," I pleaded, then went to the stern, facing away from the sport, and looked at the spray shearing off the outboard, the white foam flying away in the glitter of the blue. I felt hopeless about the pig and what I believed was unnecessary cruelty. I looked over my shoulder at the bow and saw Malé balanced there, a rope looped over his spear. He had listened to my pleading and was not going to spear the pig, only capture it.

It seemed so useless, chasing the pig and bringing him back only to slaughter it, and I said, "Kill it if you want to. I don't care, kill it."

Malé didn't answer and tried to lasso the pig. Swimming strongly and surely, the pig was almost a half mile from the island already and looked as if he could swim to the mainland. With his brown snout just above the water, flat nose awash, breath unhindered by salt sea, and his short brown legs pushing him steadily along, he reminded me much of a whale. Malé had told me that pigs could swim from island to island, but I had never believed him. After some maneuvering, the pig was caught up in the lasso and hauled up on deck, his legs tied together, his purposeful eyes now wide with fear. I turned away to watch the wake of the engine, not much caring where we were going.

For some reason we stopped just outside the reef in the shallows, and Momo got off the boat with the pig tied to a rope. Una jumped in after Momo and the pig, and Malé turned to take the boat to anchor off the *muni savusavu*, where it would be safe. In the distance, coming over the reef, were eight or nine men, brandishing spears and yelling, intent on killing the pig. Momo stumbled along the reef, the terrified pig pulling him over the sharp coral, Una running behind. Malé got furious and started yelling at the advancing men. He was fighting for the life of the pig and challenging anyone to fight him over it. This was my husband, who had taken my pleading to heart and was willing to fight for what I wanted. He was spectacular. Momo, pulled along by the tormented pig, was yelling, "*Wawa, wawa* (wait, wait)" to the men running toward him, and Malé was screaming curses.

Slowly the men retreated, and then I saw the pig break loose and run for his life along the reef, headed for the bush, men and dogs chasing him. Seru threw his spear, but it glanced off, and soon the pig and the men were gone into the bush. Malé took the boat to the sand, and we got off silently, the tears dry on my salty face, all emotion spent.

Malé wanted to find Saraqia, who had challenged him over the pig. He was ready to fight, and I followed in fear, thinking that if Malé got into a fight, Saraqia would take it to the court, and it could mean prison for him and the end of all our island dreams.

"Cool, Malé, cool," I said, following him over the hot sand, looking at the footprints of pigs and dogs, the debris, the broken shells, the refuse of the shoreline. "Cool, cool, please, Malé."

When I reached our house, Momo, Uncle Luke, Veresa, and Una were sitting on the punt drinking fruit punch. Va offered me a glass. Malé was out of sight.

Momo laughed and said, "*Bula*, Joana, you want the pig alive?

Anything you want. The pig is alive," and Momo and Uncle Luke laughed and told the story of chasing the pig.

Malé came back, and I asked him, "Malé, was I wrong?"

And he said, "I'm sorry, Joana. You were right. I shouldn't have taken the boat."

"Thank you, Malé, but was I right or wrong about the pig?"

"You were right, Joana. When I had the spear, I wanted to kill the pig, but when I looked at it and thought of our pigs, the *loloma* (love) came here," pointing to his center, "and I thought if someone killed my pig like that, I would be angry. So I was going to take the pig back to Aparosa and give it to him. If he wanted to kill it and share the meat, fine; if not, fine. When Saraqia went after the pig, I wanted to kill; it was like the old feeling, the trouble inside. I am still very, very angry. I would like to fight. Lucky Saraqia went to the other side."

Now our cousins of the *mataquali* are sitting around the *tanoa*. The rest of the men in the village have disappeared into the bush to find the pig, and all is quiet. The *yaqona* is slowly cooling Malé's temper, and the tape cassette is playing a song by Melveen Leed about wanting to go back to her little grass shack in Kealakakua, Hawaii. I am sitting in our little grass shack, comforted by the laughter of the men, my love for Malé, and the familiar singing. Uncle Veresa is sharpening a spear, Vili is setting barbs in his spear, and Toa is staring at the sand, thinking about his seed yams.

HISTORY

Galoa seems to have almost no history. I ask in vain for stories of beginnings, of the old times. The stories are fragmentary. There is the old settlement, and a few paragraphs in a book about a chief who came over here from Tavea and settled with his *mataquali* sometime around 1840. Another fragment is a story about raiding parties from Galoa going to Yaqaga about the same time. There are missionary accounts of cannibalism and brutality, detailed down to the body count. I mistrust those accounts somewhat—missionaries have a tendency to exaggerate the depravity of those they wish to heal—but I cannot be sure, although it seems that intense foreign contact has historically dislocated native cultures, encouraging violence and brutality. Certainly the introduction of muskets by the European traders helped things along in that direction.

But what of the time before all of that, when Fijians acted more purely as Fijians and when they probably didn't think of themselves as Fijians at all—as a nation—but as *taukei* (owners) of the land, the people who belong here, who are of the *yavu*, born on the mat-covered floor, and once buried there, in a corner, near the sleeping skeletons that embrace the house post.

The chiefs kept genealogies; somewhere there is a Fijian-language account of the journeys of great chiefs. But the stories that tell of the old times do not do so directly. They just point the way to a different perceptual reality, the psychic reality of another way.

Malé blends the old and new, fascinated by horsepower, covetous of nice clothes and tape cassette players and watches, yet he is often visited by the *tēvoro*. Now I know the signs: the moaning and thrashing, the muttered cursing and climactic shouting to drive him away.

One night he woke up shouting, "I am no *bokola* for you," and, turning to me, half awake and half somewhere else, told me that he had seen the *bokola* (body ready for eating)—he did not know whose—and he was fighting with the *tēvoro*; he would show him he was no easy mark. In the heavy dark of middle night, he told me that the chiefs could just say, "Get me that person to eat," and that person would be dead. If we were sitting here talking to the chief, he told me, he might order us killed and cooked; *that* was the *mana* of chiefs.

I am scratching around the edges of *mana*, trying to understand it. The answers are different. Tevita says it is a gift from the God, a gift of ability or power. Uncle Paula has the *mana* to remove a fishbone from your throat if it's caught there, as do his brothers and sisters. This *mana* was given to him by his father. I have no reason to doubt it, for Uncle Paula is truthful, a clear calm man of great inner strength.

There is a woman in a nearby village on the mainland who has the *mana* to discover the source of evil, to point to the person who is doing wrong. A few months ago sixty dollars were stolen from my purse in the *bure* on Galoa, a rather unheard of thing. It was the more damaging for it was intended as a gift to the rugby team to help out with buying the shirts; so it hurt a lot of people. Suliana wanted to take me to see the woman who could point to the thief; it wasn't necessary, for the thief was soon found out, but I would have liked to have visited the woman and felt her presence.

Long, long ago a man from a *mataquali* in Beqa, an island off the coast of Viti Levu, captured a small dwarflike creature in a deep pool. He wanted to take it back to the *koro* as a curiosity, but in return for sparing the creature's life, was given the *mana* to walk unharmed on the burning stones of the *lovo*. Now, five times a week, at the expensive hotels, the descendants of that *mataquali* walk on burning stones in the glare and pop of flashbulbs, duplicating the miracle. Uncle Navi's wife, Jokapeti, is from that *mataquali* and can walk on the stones. She assures me that it is easy, that she can give the gift to me, so that if I, or anyone else, wishes to try, we can do so unharmed. Malé says that Uncle Tevita walked the stones when the *mataquali* from

Beqa came to Lekutu not so long ago, and he saw it. It's not only a thing to show visitors; it's for family.

Malé says that *mana* is a gift from the God and the *vū* (ancestors), that with it you can do what is not ordinarily possible. The word is in the *yaqona* ceremony; it seems, like the rest of the words in the ceremony, to be archaic, without meaning other than the meaning of repeating them, of saying those words. Perhaps in that way they are truly magical, for they are calling out the unknown.

There is the *mana* in the Bible, the multiplying of loaves and fishes, the feeding of the multitude, the precious gift of food. Where does that *mana* come from? Another earlier, perhaps common, source? I cannot say they are the same, or different. The origin is lost, but the belief continues, and so does the manifestation.

An educated Fijian I know, now an archeologist in the Western world, tells of his mother's experience of hearing voices from the grave, and remembers her story that a woman in their village gave birth to two rats. He says he has no reason to believe his mother is lying; she doesn't lie.

I don't know what closing our minds to such realities does to them. Do they disappear from our experience because we cannot accept them? Malé and I lie side by side in the same bed. He sees and hears *tēvoro*. I do not. He is not lying; neither am I.

We were fishing way out near Yaqaga Point one night, and Malé decided to go farther out, toward the deep sea. We turned the boat seaward, and the gas lamp went out, leaving us in the starlit darkness. The mantle had broken, and we could not continue, so we returned to the *koro*. Why did the mantle break? Because of the pounding of the boat? Or was it a signal not to continue? I accepted the sign, was not frustrated; it was better to go back, to heed the warning. If it's easy, then go; if not, maybe someone is protecting us. The God or the *vū*? Is it the same? It is possible the world is of a whole piece. It is also possible that Fijians see it more clearly than we do, we of the education of simple and single causes.

Malé has no trouble believing in the *yalo* (spirit). He saw his uncle's *yalo* approaching him from the direction of the graveyard the night after his uncle was buried, coming into Botoi where Malé lay sleeping, waking him to ask why it was that his eldest son had not come to his funeral. Malé did not know. His uncle's *yalo* left, and Malé has not seen it since.

One night, long ago, Malé's grandfather and his children, then small, were in the bush on the mainland near the old village site near the river. Momo was there, Uncle Tevita, Uncle Malé, Aunty Sereima, and Aunty Mili. They heard terrible moaning and the sounds of fighting, the sickening crunch of club on flesh, and then the low painful cry of the *oilei*. The children were terrified, and Momo's father took them quickly through the bush to the river and told them to wait in the punt tied at the bank. As the children huddled terrified in the boat, Momo's father took his *sele levu* and walked back into the bush. He could see nothing but the darker shapes of great forest trees, but the sounds of fighting and moaning were all around him.

He slashed at the trees with his *sele* and shouted into the darkness, "What do you want? You have no business here. Come if you like, come and get me." Hacking at the bush, he shouted, "I'm not afraid of you. Come if you like, come." Abruptly the sounds stopped: only the silence of the huge trees, the distant wash of the river. He was not afraid, Malé tells me. He was never afraid of anything. He was a pastor. He had God.

The *vū* and the God can punish to teach; they can be offended because one is not following the proper way, and reach out and tug at our memories, make bad things happen to remind us of the way. They can make good things happen, can give us health and food and life. It's good to be straight, true from the center (*dodonu*). The center is not the heart; it is just above the navel, in the heart of the gut: not the mind or the heart, but the gut center, the umbilical.

From there, if we are straight and true, comes the *mana*. Momo has been given the *mana* by his father, because he was strong enough to take care of the *vanua*, the *dela ni yavu*, the source. Momo will give the *mana* and the responsibility to Malé, because he, too, is strong enough to take it.

It is a lineage, passing gut to gut, the land, the ancestors, God, man. It is not a curiosity, but a reality. It is what gives Fijians their strength and assurance.

There are *tabu* sacred places in the *koro*, places where one shouldn't go. Great piles of stones from earlier times. There is also a rock here, hidden somewhere in the bush, a rock called *bati ni namu*, the tooth of the mosquito. I read about it in a book one night, a book written by the wife of an early Yankee *bêche-de-mer* trader. She recounted the "silly superstitions" of the islanders, cited *bati ni namu*. Looking up from the book, I ask Malé.

Interest flares in his eyes. He asks, "How do you know about *bati ni namu*?"

I point to the book.

"Yes, it's here, somewhere near the point near the mainland. It is *tabu*, for if we were to go and touch it, hordes of mosquitoes would come out and plague the village. The only way they will go back is if someone from Aunty Una and Uncle Veresa's *mataquali* goes and takes *yaqona* to the rock."

I ask in vain about *bati ni namu* around the *yaqona* circle. No one seems to know the story, or no one wishes to tell it.

"Uncle Veresa, tell me about *bati ni namu.*" He says nothing. I ask Malé; he says he doesn't know. We leave the subject alone; it will come back at a later time. It takes time.

I have to still the impatience of my mind, to adjust my pace. Maybe one of these days I'll chance on more of the story, or have the rock pointed out to me. I have illusions about going to the rock with a family member, hearing the story, and maybe taking a picture of the rock and the person. The illusion comes from my European desire to know and understand.

There is another place, far down the coast, the point of Bua Bay, called Naicobocobo. It was also in the book, a dark secret valley, shadowed by huge forest trees, that was once, or still is, the jumping-off place for the spirits of the dead. I ask Malé if we can go to Naicobocobo. It's a *tabu* place, a place of *tēvoro*. He is afraid, evasive.

Malé tells me, after repeated questioning, that the *tēvoro* is the *yalo*. That is why he cannot describe the *tēvoro* to me. He tells me that only once in his life did he actually see a *tēvoro*. It was long ago, when he was a small boy living with his family on the outskirts of Suva. They were all staying in a wooden house, and his mother and father were sleeping in the bedroom. Malé, Sione, Vili, Luisa, Suli, and Ima were sleeping side by side on the floor of the parlor. Malé was sleeping in the doorway, his feet to the open door. There was another door to the outside, which was latched with a pin on the inside. Late at night Malé woke to hear deliberate footsteps slowly approaching the door on the veranda. He heard the click of the door as it opened, and then the footsteps came toward the parlor. For some reason he was not afraid, only curious. The slow footsteps came nearer, and he saw a skeleton standing in the doorway looking sightlessly at him.

Malé describes the skeleton in great detail: the ribs, the pelvis, the backbone showing through the ribs, the skull, the arm, leg, hand, and foot bones. The skeleton stood in the doorway looking at him, no more than six inches away. Then there was the soft sound of the door opening and closing and the clicking of its bony feet departing growing fainter and fainter. Malé tells me he has never told the story to his family; for some reason he never wanted to. He assures me that it was not a dream, that it was the *tēvoro*. In the morning the door to the outside was still locked with the pin on the inside.

There is talk that some people still offer *yaqona* to the *tēvoro* and by so doing are able to bring sickness and death to enemies. Sickness is often the work of malignant forces or offended *vū*. Malé's aunt in Suva has been in pain for four months now, a deep inexplicable pain in the bones of her hip and leg. She has been seen by the doctors and X-rayed, and nothing is apparent. The talk in the family is that the sickness has to do with a breach of Fijian custom. More than a year ago a distant cousin came to the family asking for their youngest daughter in marriage. The proper forms were observed, and the boy's family brought the *tabua* to seal the engagement. But the girl is young and doubtful, reluctant to get married, and the *tabua* has not yet been returned. The conclusion is that our aunt's pain is caused by this offense to tradition and will not go away until the *tabua* is returned. I ask Malé is the boy's family deliberately causing the sickness, and the answer seems opaque: not really; it's just that the *tabua* hasn't been returned.

There are stories and stories: the sand turns into a boat, takes Uncle Veresa to Vanua Yalewa. *Tēvoro* live in the *baka* trees, haunt the *koro*. It is not safe to take infants around in the dark. One of Malé's uncles was killed by a huge barracuda, slashed to pieces because he had been offering *yaqona* to the shark. There is an uncle on the mainland who has the *mana* to give a woman a baby if she can't conceive. I want a baby. Do I dare believe?

The women in a village up the coast past Labasa go into the bush, adorn themselves with leaves, perform a *meke* to summon the rain. It rains. There is a pool in the Yasawa Islands that is reached by a narrow opening. It is said that no matter how big the woman may be who wishes to bathe in the pool, she can enter through the opening, but if she is pregnant, then no matter how small she is, she cannot squeeze through into the pool.

Naigani is an island where if you catch the *saqa* and throw the bones back into the sea, the bones will swim away, becoming *saqa* again. This is a vast ground whereon lie diamonds of a different glittering reality. I think it may be reached through the language and motion of the *meke*. The language is so difficult, so complicated in meaning. The word *sau* has more than twelve dictionary meanings: to repay, to answer a question, to cut reeds and bamboo, the reeds themselves, the act of placing them, a high chief, the orders of a chief, a way to break something, to let down the net in fishing for turtle, to think of, to cast a spell, to sail in forbidden months, to clap the hands together lengthwise, outside, the white cowry shells tied to the *tanoa*, a bird—and more that are hidden between those meanings. The language reflects the reality. It is multifaceted, sparkles like the diamonds on the sand.

⯭ ⯭ ⯭

In the dark stifling night, Malé's *tēvoro* comes out. I wake under the mosquito net with a sharp pain in my side and ask Malé to massage it for me. He tells me that the pain is because I yell at him. His *tēvoro* is causing the pain.

"It is because when you yell I only laugh at you; I don't hit you. If I hit you, I would kill you, so I just wait. I know that something bad will happen to you."

This is better than silence. At least Malé is talking about what upsets him. I want to know more; I calm myself to listen.

"Is it like your aunty's pain?"

"Yes."

"Because of the Fijian thing, because I am not acting like a Fijian?"

Silence.

"You have had a pain in your stomach for a year or more. Is it because you are doing something wrong?"

"No, my pain is different. But your pain is from my *tēvoro*. So I just wait."

A low quiet wave comes in, cleans my spirit, a release.

"Do you really hate me?"

"Yes."

"Did you ever love me?"

No answer.

I go out and lie under the *baka* tree, the dark leaves strain the starlight, glints of distance shimmer among the leaves. This is something I can understand. Hatred is an easier emotion than love, and Malé's *tēvoro* an honest participant in our relationship. Paradoxically it brings us closer. Hatred in the air is harmless. Exposure kills it. Hatred can live only in the remote corners of the soul, feeding on itself. We are finally beginning to communicate. We are bridging a gap deeper than differences in language, a deep-sea trench that lies between our cultures.

We are trying to construct a marriage. A delicate structure wrought in time. We can harmonize our cultural differences if we can appreciate them. Culture is a garment that clothes the soul. We may never be able, or even want, to exchange our cloaks, but what matters is the perception of each other's realities, even if the reality is hatred. I know the hatred will pass and come again, so will love; it is inevitable. I do not fear Malé's *tēvoro*, only his remoteness. A tiny corner has been lifted, a passageway I wish to follow.

A while ago, bewildered by our failures, I asked Malé's brother-in-law Nacā to help me try to understand the Fijian man. He told me that nothing is hidden; the emotions flow freely; there is no evasion. But he also told me that the head of man is sacred, *tabu*, and that it would be years before Malé yielded his head to me. You must change, said Nacā. You have the experience, the education, the ability to change. Malé cannot change. He is of the village, of the *vanua*, and if he were to change, he would be lost, neither European nor Fijian, robbed of his strength, of his way.

I know that I cannot go back; I have opened something that I must complete; I can no longer go back to my old ways, the easy superficial satisfactions of status, the fancy food and clothes that soon become boring, the repetitive relationships that I flee from as soon as they become difficult or demanding. I know that I must learn to replace my huffy pridefulness when someone criticizes me or tells me what to do with a more substantial sense of my own dignity and value.

I also know that I love Malé, that I respect and appreciate his character, the self-confidence that derives from his being secure in his culture and not questioning it. Ever since my early education in anthropology led me into questioning my own cultural values, into seeing them as one cluster of many possible responses to the problem of figuring out how to live gracefully through human life, I have been confused and unstable.

Fantasy was my response, and now I have led myself to the testing ground of my fantasies. It is unlikely that I can ever believe in fantasy again.

TIMOCE

Timoce lives in a big old reed house with a corrugated iron roof, filled with furniture he never seems to use. There are two wooden beds with mosquito-net framing, a number of wooden chairs, and two or three tables hidden in the gloom. Whenever I have gone to Timoce's house, he, like everyone else, sits on the mats, the furniture piled with suitcases containing clothing, neatly folded mats, stacks of pillows, and household articles. Timoce's house is a dim reminder of what was once a high art. The old reed walls, now sagging and askew, were once fashioned in intricate decorative patterns characteristic of the finest skill in Fijian house building. It is the oldest house in the *koro*, iron roof aslant, settled on its sand-and-rock foundation like a tired hen. I see Timoce in the garden, wifeless and alone, a weary figure, most often silent at *yaqona* sessions, like the few other old men in the village who sit in the places of honor opposite the *tanoa*, smoking and coughing, murmuring *oilei* as the discomfort of the long sessions creeps into their bones.

Yesterday I hunted up Turaga, Malé's cousin who speaks English, and asked him to come with me on a story quest, having been advised by Uncle Veresa that Timoce knew the story of *bati ni namu*. Notebook in hand, I followed Raga, who approached Timoce's house in the customary quiet polite way, standing offside the doorway, for one doesn't stand in the doorway, and inquired if Uncle Timoce could *talanoa* (talk story) for a little while.

"Yes, come in," he said, and we entered the gloom. For a

change Timoce sat in a wooden chair in the corner, the soft brown light falling on his Sunday beige shirt and tailored brown *sulu.* Raga and I sat on the mat, and I listened while Timoce talked and Raga interpreted. I watched the play of light on Timoce's weathered face, the dark patch of skin under one eye, reminding me of the stories I had heard of men blackening their faces to look fierce in war.

"*Bati ni namu? Ro Bati Namu* is the name of the *tēvoro* that guards that place. It was in 1921; there were plenty of mosquitoes, mosquitoes so thick you can't sleep, even under the net. Everywhere you go there were mosquitoes," Timoce drew his hand through an older air, "so many that when you move your hand you touch hundreds of mosquitoes. The people had to bury themselves in the sand, dig a hole, and crawl into it, with just their noses outside to breathe. A wave of mosquitoes, just like the locust. At that time the people in every house had a *tobu ni madrai*, a hole where they put the cassava to age before making Fijian bread, a fermentation pit. The people thought the mosquitoes were coming from the pits, so they filled in all of their pits; that's what the people thought. That's why no more *tobu ni madrai* on Galoa.

"Long ago some guests came to Galoa, people the people here didn't like very much. Maybe they wanted to come here to find a wife or something not good. Only one person from Galoa was to go to *Ro da Cewa* (the place where is *bati ni namu*). They go, and they take one stone from the *koro* to the *vatu (bati ni namu)*, and have to say, 'Let the mosquito bite the guest.' Then when the guest comes to the grog party, he can't sit still, only bit by mosquitoes; they land on his face and hands and in his eyes and mouth. When they hate the guest, that's what they do; they didn't fight or anything; they just let the mosquito bite. When the guest leaves, they still have the mosquitoes, so the *tolo* for the *vatu* takes the rock back from the *vatu* and says, 'I'm taking the rock back'—that's what the *tolo* says, 'I'm taking the rock back.' "

I interrupt, "The mosquitoes go back in the *vatu*?"

"Yes, then the mosquitoes go back."

"Who is the *tolo*?"

"The *tolo* is the oldest man from the Cake *mataquali*."

"Are you the *tolo*?" I ask.

"Yes," says Timoce, "I'm the *tolo*."

"Can you show me the rock?"

"No, only the *mataquali* can go there."

"Who is in the Cake *mataquali*?" I ask.

Raga tells me, "Timoce, Uncle Veresa, Uncle Saimone, Uncle Jo, Aunty Una—that's the Cake *mataquali*."

The story is over; Timoce wants to go drink grog.

"We go *somi qona*?" I ask.

"Yes."

"Can come back some other time, *talanoa*?"

"Yes."

"Thank you, thank you very much for the *talanoa*."

Timoce waits in his chair in front of the sagging reed wall, looks out the doorway, smiles, "It's finished now; we go."

Outside the men are gathered in Uncle Saimone's *vatu nu loa*. Malé is pounding the grog, Toa and Vili crouched next to him joking. I leave Raga with them, walk back to our *bure* holding my notebook. That was long ago; it's finished. Nobody goes to the *vatu* anymore with a stone to get rid of the guests.

MATA NI GASAU

I had to go to Hawaii. It was going to be a brief visit, but I had to see my mother, who could meet me on Maui, but who hadn't the strength to make the long journey to Fiji. I could not bring myself to travel the great cultural distance between Los Angeles and Galoa, so we agreed on a mid-Pacific meeting point as our compromise. My son, Ian, would accompany my mother on the plane, and we would all have a week together in a hotel, before dispersing again to our separate lives.

I was fearful; travel makes me fearful, and there was the sense of leaving to go back to something I didn't want to get trapped in. Malé and I were on Vedrala, sitting on the wet sand underneath the mangrove clumps, watching a big school of fish clanging around in the shallows, and I voiced my fear to Malé. I was looking for a form of blessing, something that could protect me on my journey, and was also looking for some kind of absolution.

"You know, Malé, I know I haven't been all that good to your family in spite of all the things I have bought. I know I have been angry and resentful and have over the year said and done a lot of things that have hurt Momo and Nei and the girls. Is there a Fijian way I can ask for forgiveness before I leave, some way to make it right before I go?"

Malé was pleased. It pleases him when I want to follow the Fijian way, when I show interest and respect. "Yes. You buy the *yaqona* for my father and tell him it is your *mata ni gasau.*"

Malé taught me a short speech in Fijian, my first such attempt, and I wrote it down and struggled to memorize the short unfamiliar sentences.

The *gasau* is the tall reed that is used for thatching the roof of the *bures*. It is strong and durable, but dangerous to handle. It is razor sharp, and unless one is very careful while handling or walking through it, deep and painful cuts result. The *mata ni gasau* is a ceremony designed to clear the way of hidden dangers, and we would do it to clear my way and bring me safely home.

Three days later we gathered in the family eating house just after dinner. I was nervous about trying to speak Fijian ceremonially before the family. It was only three sentences, but the words still didn't make enough sense to me to guarantee that I would repeat them correctly. I reached behind me and brought out the bundle of *yaqona*, wrapped in newspaper and tied neatly with a string, leaned forward toward Momo, and patted the yaqona like I had seen him do so often.

"*Momo, i'ai e oqo mata ni gasau vei i'o. Ni bere niu gole i Hawaii, me ua ni dua leqa i oqui iakolako. Au varogo taina vei i'o, sa oya ga ana ena levu, vina'a va'a levu.*" (Momo, here is my *mata ni gasau* for you. Before I leave for Hawaii, let there be not one problem in my journey. I have spoken to you, this is all I have to say. Thank you very much.)

I looked into Momo's tired eyes, shining with amused tolerance at my stumbling speech. I was sure I had left out some vital words. Aunty Mili nodded at me and smiled, thank you, thank you. The rest of the family grinned. Momo answered, the liquid Fijian rolled r's softening my pride. He accepted and prayed for my safety, for the safety and well-being of everyone involved in the journey: for my mother and son, who would travel from the mainland and return; for Malé, who would accompany me to Nadi and return; and for myself, who would travel and return—all in good health, all safely, without any *leqa*. I felt calmly reassured in the knowledge that whatever could be done had been done, and it was for me and my family a smooth journey, unmarked by problems.

After the ceremony we sat around the *tanoa* drinking *qona* and telling stories. Since I was leaving, I had the family's attention and could ask about things that were of interest to me, and perhaps find answers. Tomorrow, after I left, the family would also gather and drink *qona*, to follow in my footsteps, so to speak, to accompany me on my journey. But tonight we were all together, and I wanted Momo to talk about the old ways, as much for his children's sake as for my own, so that we would all know and could continue the story.

Somehow it is not the facts—the details—of the past that are important; that is history and is lifeless. It is the way that is important. If the way is followed, then everything fits together, goes smoothly; the past continues into the present, is not exotic, is informing. The way has to do with manners, with each person's being an easy interlocking piece of the picture, not trying to be the whole picture. There is a way for small Malé and small Vili; it changes for small Jone, becomes more responsible. The first wood one gathers, the first yam one digs, all are important to the unfolding of a life that fits the puzzle of the family, the continuity of the *mataquali*. Qare's young daughter Selai picks up the spent coconut shells, straddles the grating board, and grinds the almost-empty shells. She flows smoothly into the pattern, later will carry small bundles for her mother, much later will dig clams and dive for shells, catch fish on a line, make fires and cook food, bear and nurse children, answer questions, and carefully save her daughter's first bundle of firewood, save it for a small family ceremony, a *magiti*, to celebrate the fact of her contribution.

The things of the Europeans are lovely, but they are things. Here, if the engine breaks down, we can still paddle to the fishing grounds. The fish are the gift, the food that gives us life. The engine makes it easy; the fish make it possible. It is better to listen to the old people, not only for the story, but because the act of listening shows you have respect, not only for the person, but for the way that person lived.

When you change things around too much, you break them.

We were not always the Nalomolomo *mataquali*, Momo tells me, we were once the Wasa *mataquali*, and the Wasa lands were ours. Then the government made the law to register *mataquali* lands so there would be a record, and a man named Nepote was the only one in the *koro* who could read and write English, so he was the one who wrote down the names of the members of the *mataquali* in the government book. He changed the names of the *mataquali* and of their lands, changed them for his own reasons. After he did that, things started to go wrong.

Now the *mataquali* are nothing; they don't do what they used to do to make the village work easily. Everyone goes to the village meeting and talks and talks and then doesn't do anything about what they say. They always have big plans; then they don't do anything; they just like to talk. There was a cooperative store for the village. Someone stole all the money, and the store went broke, had a big debt that it couldn't pay. Nepote's son went mad (*lialia*). The village bought a boat from the National Development Bank, a good boat to go fishing and to take people from the village around. Nobody took care of the boat; nobody could decide who was the captain. The engine broke; the boat sat on the mud in the river for a long, long time; finally the bank took it back, rotten and rusty. They started a store to serve the *koro*. There is no store. Look at Nepote's grandson; he is half mad, is trying to eat himself to death. Everyone goes his own way now, says Momo.

When forms are broken, great trouble results. The grass is sharp. It grows in head-thigh thickets. A blade can slice your eye if you do not watch where you are going. The *gasau* is useful and dangerous; care must be taken. This life is a journey through known and hidden dangers. It is well to keep to the old forms. We do not know what the God will do.

ONLY THE GOD KNOWS

It was a Sunday morning last September. Malé and I were in bed, not fully awake. Nei, my mother-in-law, rustled at the door thatch and crawled into our *bure* bearing a bundle of clothes. She sat her heavy body on the mat, and folding the clothes, started talking with Malé in Fijian.

"What does she want?" I asked Malé, who ignored me and talked with his mother. Still half asleep, I took the pile of neatly folded clean clothes she offered. Her face looked worn and unhappy, and after a few more words, she crawled back across the mat and left.

I got out of bed and, tying a *sulu* around me, walked the thirty feet across the compound to the kitchen and got two cups of hot water for coffee. I could hear the rhythmic beat of the *lali* sounding the call to church, a peculiar intense rhythm in which the left hand marks the time and the right counterpoints. Shortly after the sounding of the *lali* came the clank clank of iron being beaten on iron, the village's approximation of a church bell, and soon after that, the liquid voices of the choir. Sunday morning church had begun. The coffee tasted good, thick with instant milk and brown sugar, and the singing of the choir soothed me, as it does so often. Suddenly Una's face appeared in the doorway. She talked fast and looked worried.

"Something's happened to my mother," Malé said.

When I reached Botoi a few minutes later, Nei was lying on the bed. Momo, Malé, and Una sat looking into an uncertain future.

"Nei, *a cava*? (what is it)," I asked, bending over her.

She looked at me through filmy eyes, ravaged from bending over smoky cooking fires, "*Maqa* (nothing)."

I asked Una what had happened. Nei had fainted, passed out,

and then came *lialia* (foolish); she had talked senselessly about preparing some food for a nonexistent *magiti*.

Momo looked at me brightly with youthful hope, "*A cava*, Joana?"

"I don't know," I said, thinking that there was no way to know, but also thinking that she had had a stroke, a small stroke. She seemed all right, was not paralyzed, but weak and far away. My mother-in-law had been sick for years with high blood pressure and diabetes, worked constantly from dawn to midnight, and had borne and nursed ten children. She was worn out.

Two hours later, when the tide came up, we got into the boat to go to the health center in Lekutu, where there was a paramedic. Nei was weak, but seemed all right. We just go to check; it's good to go check, we told one another cheerfully. We made a bed for her in the boat—pillows and a mat—and Suli packed a small basket with her and Nei's belongings, in case they had to stay overnight. Momo joked with that bright intensity he uses when he is trying to be strong. Malé drove the boat and looked vacant; Suli played with small Malé; and Nei looked out the salt-stained window of the boat at a blurry, receding island, an island she would never see again.

The banks of the river were drying. Once-lush rain trees showed gaunt branches, their silken flowers a memory of summer. The tide was still low, and we traveled slowly, careful of the shallow places, making small talk, anxious, but not yet fearful. We tied the boat near the Peace Corps fish market, and Nei got out to sit on a thick, twisted root in the shade, catching her breath. She looked really weak, and I asked Malé if he and his father would carry her up the path to the health center.

"No. It's all right. My father went for the van." But there was no van, and we sat for a while waiting. Finally, in response to a question from Momo, Nei pulled herself up heavily and began the slow step-by-step ascent of the path. Neither Malé nor his father offered her his arm in support, walking instead slowly behind her as she made her way.

"Malé, carry her, she's really tired."

"No, it's all right. She can walk." Malé and Momo were trying to give her strength by pretending that she was strong. It was the Fijian way. And so we went step by step up the path, across the gravel road, and up the small hill to the Vale ni Bula, house of health and life.

Her blood pressure was alarmingly high. Given some medicine, she was put to bed in the makeshift hospital: eight iron beds and not much else. I went the short distance down the road to Munci's store to charge some food for her and Suli, who would stay with her—two cups and plates and spoons, some mosquito punk, a water bottle, and matches—and came back. There wasn't much to say or do. We stood around the bed looking at a woman who wanted only to sleep, and who could do little else; and Momo prayed in his strong resolute voice, a long prayer, over his sleeping wife. Suli and small Malé stayed with Nei; and Malé, Momo, and I returned to the *koro*.

The days went by, days of little change. Momo went to see Nei daily, sometimes accompanied by Una or Ima or Va. She was never left alone. There were some days of hopefulness, when Una or Suli would proudly assure us that she was eating well, detailing the exact amount of tea, biscuits, eggs, oatmeal, as if that could stay the slowly creeping flood.

Malé was silent and terrified, wanted only to drink *yaqona*, joke with his cousins, and sleep. When he visited her, he sat by her side looking at her, saying nothing.

"What if my mother dies, what if she dies?" He finally voiced the unanswerable question one soft star-kissed night.

"She won't die. Plenty of people have high blood pressure,

and then the pressure goes down and they get better," I tried to reassure him. But in his deep Fijian way he knew, and so did Momo, who told me that if the God wants to take her, it's good. Never, when visiting, did I see them touch her. They stood silently by the side of the bed gazing at her receding soul.

Three weeks later Nei had lost control of her bladder. I hired a van, and we crowded in: Malé, Momo, Suli, small Malé, Nei, and I. We took the long hot dusty ride to Labasa to have her checked at the hospital. I rented a hotel room for Malé and me; Suli, Momo, and Nei would stay with an aunty, and we all came to the hotel room to rest before her appointment at the hospital. Momo was entranced by the room; he bathed and ambled around, investigating hotel exotica. Nei lay on the bed, ate a soft-boiled egg we ordered from the kitchen, and in some small corner of herself seemed happy. It was the first time she had been inside a hotel. Malé showed off his hotel knowledge, and Suli chattered. Small Malé was afraid.

At the outpatient clinic the doctor fired questions at Nei in a high irritable voice. She could not understand his Indian-accented English, could not understand any English, and I interpreted for Suli, who asked Nei. She was too tired to answer much, and it seemed as if the doctor was interested only in blaming the paramedic at the health center for giving her the wrong medicine. Anyway, it was late; he said he should have gone home by then, so after we had waited three hours to see him, the doctor told us to come back the next day.

I rented another hotel room for Nei and Suli, and Momo went back to Galoa in the van with small Malé to bring back the forgotten mats and mosquito net and blankets that Nei and Suli would need in Labasa.

To get our minds off our fear and worry, Malé and I went out dancing that night at a seedy place, dark and menacing. Malé sat me at the bar and told me to stay there and disappeared into the gloom. It was noisy and hot, and I went out onto a small balcony to get some air and was instantly surrounded by a group of drunk aggressive men. Ducking my way through I went back to the bar, where Malé was furious at me for having moved from where he had left me. He picked a fight, pushed me in the head, and told me not to leave his side. It was dangerous; words were no use; he was drunk and panicky about his mother. I went down to the street to calm my temper. Jo, a cousin from Galoa, instantly appeared from upstairs to protect me on the street. I told him I was going back to the hotel, and he went up to get Malé. The three of us left, walked in silence down the near-empty street to another, more-respectable place. There Malé drank and talked with his cousins while I danced with anyone who asked me.

The next day at the hospital brought no relief. I had expected that they would admit Nei, but she was told to go home, take medicine, and come the next week for another test. We left Nei lying on a mat in the disorderly yard of Aunt Seiniana's tiny house on the outskirts of Labasa and came home. Momo and Suli stayed to care for her. Nei was taking her medicine. She was okay. Malé worried, and I tried to reassure him. A week and a half later, he wanted to go see his mother; something told him it was bad, and so he and Ima and I returned.

Nei had been admitted to the hospital. She couldn't walk, but was still eating and could talk. Mostly she slept, raising tired eyes to barely acknowledge greetings. Momo ran around like a madman, walked the two-mile walk to the hospital at three in the morning to look at his diminishing wife. He couldn't sleep; he paced, made plans, prayed. Va had come from the *koro*; she, Suli, Ima, Aunty Seiniana, and Aunty Vulase each took seven-hour shifts at the hospital, sleeping on the floor, in constant attendance. Two days later, Momo, Malé, and I made plans for her funeral. It was clear that Nei would die.

I went out in search of a coffin. I climbed the wide dusty stairs in back of a hardware store to look at boxes, small narrow things; it hardly seemed possible that Nei's big body could be squeezed inside. Flies buzzed in the oppressive heat of the attic. The boxes were seventy-five dollars and had handles. Yes, she will fit inside; don't worry. I felt distant and calmer. It was easier for me to cope with details than to stand unspeaking with Momo and Malé, waiting.

The next afternoon she died. She slipped quietly out of her tired body; all systems shut down slowly, feeling and life leaving, to flicker for a moment in her tired eyes, and then go out. We all gathered around her bed in the hospital—uncles, aunts, cousins—and Momo stood next to her and prayed. When the prayer was finished, he turned back the sheet and patted her sagging belly, a belly that had borne his relentless seed. It was the first time I ever saw him touch her. Not once since the first day she fell ill was she ever left alone. There was always someone in the family with her. She was alone now.

⁂ ⁂ ⁂

That night we gather at Aunty Seiniana's house. Mats are spread on the grass and the *tanoa* brought out. Thump, thump, thump—the *ta bili* pounds the *waka* into powder. We are listening to Radio Fiji, waiting for the news of Nei's death to be broadcast to summon the relatives from Suva, Taveuni, Lautoka, and Labasa to the funeral. The harsh crackle of the radio is heard against the deep earth pounding of the *yaqona*. Momo is exhausted and strong, making plans with his brothers and cousins. Vaseva, heavy with her first child, at seventeen a child herself, is in shock. Malé is quiet, strained. Today we bought him sunglasses so that no one could see his tears. A van arrives in the darkness; headlights flare. Silent dark shapes of people emerge from the van, the women carrying rolled mats and *sulu*, the men cartons of food—sugar, tea, flour. The women go into the house with the mats, pile them on the floor before Suli and Aunty Seiniana, and cry the *oilei*. The men sit outside in the shadows, drink *yaqona*, and divide the responsibilities for the funeral. Who will pay for the vans to bring the coffin and the mourners? Who will buy the *yaqona*? Malé is dry eyed behind his sunglasses. He is very far away. We wait in vain for the radio announcement.

The next day is a frenzy of shopping. Bags of flour, sugar, rice, other food, and formal *sulu* for Momo and Malé, white shirts, a nice dress for Vaseva to wear to her mother's funeral, a dress for Banoko, sandals for Momo and Malé. I want everyone to look nice, to have at least the pride of handsomeness in grief. At the end of the day, we pile into the van; the rest of the family will follow tomorrow. We reach Galoa with a flurry of things. Vili is waiting, and we load them onto the boat in the darkness. Possessions that cannot fill the emptiness. We drink tea and sleep.

Malé wakes me early with a kiss, leaves to clean the fish that Tovi, Vili, Qare, and Toa caught last night. A blessing: six big *walu*, three bundles of *bonito*, *oqo*—more than sixty kilos of fish—so no need yet to kill the pig. I sleep again and wake to the *oilei*. Ima sits on the punt, rocking herself like a child. She is screaming, "*Oilei na, oilei ta*." The sound freezes my marrow. I cannot cry. Una, Ima, and Suli are screaming their sorrow, their voices cracking and harsh. Momo is red eyed, frantically busy. Malé is working, working, cleaning fish, cutting firewood, building the *vatu nu loa* for the *magiti*.

Malé stops working for a few moments, comes into the *bure*, and we drink tea, share a smoke. He tells me he wants to see his mother's *yalo*, her spirit. Tells me again the story of seeing his uncle's *yalo*, Toa's father, come from the graveyard.

"Joana, I hope I can see my mother again, just once. Just once I want to see her *yalo*."

"Maybe, Malé, maybe you can; maybe she will come to you."

The *oilei* lapses. It is quiet, hot; doves are cooing, mynas chattering, children's voices raised in laughter. I think of Nei lying on the hospital bed, slowly losing her senses—sight, taste, touch—slowly diminishing. I meditate on the pleasures of life, the feel of things, the taste of food, of drink, the breath of wind, sun warming tired bones, water washing dirty skin. I tell myself that I must remember not to let my mind defeat my body, not to become depressed and take lightly the gift of life and health.

A boat comes. Luisa and Nacā have come from Suva. They are dressed in black, with finely woven mats tied around their waists in the Togan fashion. Luisa carries a huge pile of exquisite mats and *tapa*, kneels and arranges them in the shade of Vili's *bure*, the big ones on the bottom, the smaller on top, creating a moment of shimmering beauty, the bright wool fringes and finely worked patterns etched in the shadows. Luisa is solid, calm, greets me kindly, explains that since Nacā comes from a Togan lineage, they will do this in the Togan fashion. Nacā, Luisa, and Momo sit in the center of the mats. Nacā presents a *tabua* to Momo, to reintroduce his wife back to the *koro* of her birth for her mother's funeral. Then the mats are carried into Botoi, stripped now of beds and furniture and *tapa*, stripped to become the *vale ni mate*, house of death. Momo puts certain *tabu* on the house. No one is to cross directly in front of the doorways, but to pass in the back of the house. There will be no joking or laughing inside. The women go inside to continue the *oilei*, to receive the other women with their gifts of mats and *tapa* and *sulu*. The men are outside, digging the *lovo* for the pigs, cassava, *dalo*, and yams, and peeling and preparing the cassava for the *lovo*. The thump, thump, thump of the *ta bili* pounding the grog is heard continuously.

Malé and I bathe in the *vale ni sili* on the beach and dress. It is time to take the boat to Nakadrudru to pick up Nei's body. Malé is wearing his sunglasses. He is lost. We smoke and say nothing. The tide is high. We reach Nakadrudru in fifteen minutes. Waiting on the banks of the river are perhaps fifty silent people standing in the dappled shadows: women and young girls in black, men in formal *sulu*, people I have never seen. The plain plywood coffin has been painted a rich red-brown, stenciled in white with *tapa* patterns. It is banked with masses of brilliant red and purple bougainvillea and ferns and is laid carefully on a new mat in the boat in the same place where Nei lay when she made her last journey up the river. A number of people crowd in and sit around the coffin. Malé drives slowly down the river, followed by three other boats carrying mourners.

One green mango, suspended in time, hangs over the water. Green reflections slide by. When we reach the shore, Malé tells me to stay on the boat. Before we get off the boat there will be a presentation of a *tabua* by the chief of Galoa to welcome Nei back to the island of her birth. A relative of Nei's, a man I do not know, is riding on the bow in front of me. He will receive the *tabua* on behalf of Nei's family. A silent crowd is waiting on the shoreline. The chief steps forward into the water and, holding the *tabua* out in front of him, stretching taut the woven sennit string, speaks the now-familiar words. The ancient formula rolls out, "*Ē Ō dua sa dua sa. Ah! Ah! muduō, ai, mana*. Words that comfort, calm, "Yes, it is so, a gift."

Oilei na, oilei ta. Ima's voice is harsh with weeping. Vaseva looks into the darkness. The coffin is carried past the eating house, past Vili's *bure* and the *vatu nu loa* into Botoi. The *oilei* inside intensifies. I shiver.

Malé stays up all night drinking grog, digging the *lovo*, finding work. There is a strain between us. I want to communicate with him, share his grief. He shuts me out, does not speak, and we argue in the morning about using the *Dolphin* to go to the mainland to pick up the cow that must form part of the funeral ceremonies. The argument delays him, and he leaves late, in anger. I watch the boat roaring off and listen to the rise and fall of the *oilei*.

I am worried about Vaseva; of all I think she is suffering the most. She has been the kindest, the most gentle, with an innate sense of beauty and compassion. She is still a child and has just lost her mother, and is carrying her first child, with no one standing between her and tomorrow. By late morning Malé has not yet returned from the mainland with the cow.

"Come, Vaseva, we go to Nabau and bathe." Small Malé, small Suli, Va, and I go down the path into the silent peace of the bush, past the village graveyard, fragrant with fallen frangipani blossoms; under the row of great old mangoes, heavy with ripening fruit; and under the coconuts to the cool sanctuary of the well. Silver water splashes into the galvanized bucket. I wash the children first and then myself, grateful for the coolness and the scent of soap. We linger in the glade, the thick leaves of the cocoa trees a canopy against the harsh sun. Two heavy phallic stones guard the rock pile of the well. Lizards flash their heraldic brightness in shafts of sunlight. Big fat frogs plop like falling fruit in the thick brown leaves.

Vaseva and I dress in our *bure*, her new pink embroidered dress a flag of innocence. I am keeping her close.

Outside, the family has gathered in front of Botoi to go into the *Lotu*. Momo calls me over to join the procession. The *lali* sounds, the deep beat of the *lali*, answered by the high metallic clang of the bell, slow and mournful, extending the silence. We file into the church. Malé has still not returned.

The minister from Lekutu who married us in this church leads the prayer. I look up at the stained-glass window, bright and jewellike in the sunlight, look at the coffin banked with its now-wilting blossoms. Nacā, who is a minister himself, mounts the pulpit and delivers a sermon. Nothing happens. No feeling stirs the air. We are suspended in time, like the green mango, waiting to ripen.

Then Momo rises, wearing his new white shirt and pale blue *sulu*. His eyes are brilliant, and he begins to speak, smiling and laughing. God enters the church and surrounds us with his awesome presence. Momo pounds the coffin, spreads his big hands over the thin wood that separates him from his wife, tells stories. He tells the story of the photograph that Nei wanted taken a week before she died. How she asked Momo and small Jone to come with her into Labasa and pose for the picture, standing stiffly in front of a painted backdrop of a pillar, a distant sea and sky. Momo tells how Nei knew she was dying; that's why she wanted the picture. They never had a picture of the two of them together in all those years; only then was there a picture. Momo laughs, pounds the coffin again, raises his arms to the heavens; she knew, it's good, it's all good. Only the God knows, only the God is supposed to know. Momo's aura fills the *Lotu*. He is intimate with his wife in the coffin, is unafraid, laughs and shouts, is one with the crazy intensity of life and death.

A hymn slides the afternoon into silence. We follow the coffin outside, across the *koro*, past Botoi, and down the sand toward Wasa. It is low tide, and the procession strings out, picking its way through the rocks and corals. In the distance, heading from the mainland, I see the *Dolphin* roaring toward us with the dead cow.

A grave has been dug in the dirt back of the shore at Wasa. The grave is lined with new mats; the coffin is already inside when Va and I reach the graveyard. We sit on the ground in the shade of the great *baka*, and Nacā reads from the Bible. The women are silent; the *oilei* is finished. I look into the faces of the family: Momo, Sione, Luisa, Vili, Ima, Una, Va, Banoko, Boko, Suli. Small Jone sits alone on the dirt, a patch of sunlight blazing his white shirt, molding his delicate features. A cousin jumps into the shallow grave, wraps the coffin tightly in the mats, carefully folding the corners into a neat bundle.

The first shovelful of dirt falls on the coffin with a dull thud. Soon the grave is filled. Momo runs around picking up heavy stones to make a rock wall around the grave. Men and boys go in and out of the bush, returning with stones. Momo orders the arrangement in an almost hysterical voice, rushing around rearranging rocks. The wall is filled in with sand, and first the men, then the women, pose behind the new grave while I take pictures. Soon everyone is gone. There is only the unadorned grave and Vaseva and myself left at Wasa. We go and sit on the beach in the shade of the *baka*, some distance apart, each with our own thoughts. I look at the sky, clear brilliant blue. One white cloud, vast and pure, detaches itself from the mainland and floats up into the sky.

Back in the *koro* Malé has returned with three cows, their legs tangled together, bellies slit, eyes glassy and staring. He is furious with me that he missed his mother's funeral, blames me for making him late with the boat. He is right; I have little to say, but think it must have been God's will that he did not see his mother buried. Malé goes to the *vatu nu loa* with Vili, Qare, and Toa to cut up and portion the meat. This is the *burua*, the sharing of a freshly killed animal with the members of the family, that must form a part of any funeral. The men work hard in the hot sun cutting up the three cows and handing out the meat. The people who are leaving today will take their meat home with them; the ones remaining in the *koro* will cook and eat it here. Malé has not slept in three days.

Now that the funeral is over, the *tabu* are lifted. A *yaqona* ceremony is held in Botoi to open the house of death, and one is held to cleanse the women who have been sitting in wake around the body. It is symbolic and actual: when the ceremony is over they go off to bathe, change their clothes, and reappear free of the *tabu*. Mats are spread in the old *Lotu* and the great *tanoa* brought inside. There, at the scene of our wedding feast, we gather to bring the blessing of life back into our lives. Late at night the guitars and ukuleles are brought out, and at Momo's request, the singing begins. Malé, Vili, and Qare sit together in a close circle, facing one another, and begin to sing. All of their sorrow is in the music. It is the most beautiful I have ever heard, their voices crying into the unknown, singing of love and loss and life, mourning the death of their mother. Malé's voice, that never sounded the *oilei*, now sounds the perfect harmony of grief.

⁂ ⁂ ⁂

In the morning, after another sleepless night, we fought. It was all too much for me—the last weeks, Nei's death, the constant pressure of people, no time for reflection, for gathering my own spirit. I forced myself to pack my things; I had decided I would leave Malé for a while to his *yaqona*, to his cousins, to his remoteness. Malé watched, listened to my ranting, was frightened.

"If you want to go, you can go, but I'm asking you to stay. Please. Marriage is sometimes really bad, but sometimes it's really beautiful."

JOANA

At the top of Seatura, the great volcano that dominates Bua, is said to be a lake so crystal clear that you can see the sand on the bottom; but if you throw a stone into the lake, the stone will never reach the bottom. The old geography of Fiji does not mention the crystal lake of Seatura, only its precipitous dark slopes furred with forest trees that now cloak the once-fiery caldera.

It is a familiar presence. The long low flanks of Seatura remind me of Mauna Loa, where I once wandered as an acolyte, seeking Pacific mysteries. I sit on the sand under the great *baka* at Wasa looking across a pastel choppy summer sea at the mountain, sit companionably with the ghosts, for they, as always, are more comforting than men. Behind me, in the sloping natural amphitheater that forms the basin of Wasa, the leaning *gasau*, the quivering palms with their brilliant rust trunks leaning seaward, the thick green spiky leaves of the *voivoi*, and the dense hot smell of fecund summer recall my first walk on Kilauea, where I felt strongly that I must somehow be visiting a botanical garden, an arranged exhibit, so fantastic was the vegetation. Now the reality of these southern tropics seeps into my consciousness, slowly replacing my sense of constant discomfort with its awesome and forbidding beauty, its lush inaccessibility, the remote fertility of this creation.

Another person is on the earth. A new person comes into the

household, nurtured in Vaseva's womb, the issue of an impossible desire for a love that never was, for a father that has never showed his face. She is named Joana McIntyre Varawa. *Noqu yaca* (my namesake) sleeps in a nest of pink pillows under a lace-trimmed mosquito net, her small dark face relaxed, her tiny fists curled around an unknown future. She is calm and lovely. She is at the beginning, surrounded by love, knowing only the warm milk from Va's soft breasts, her mother's gentle touch, and the atmosphere of admiring eyes. She sleeps, wakes, nurses, sleeps again, to the mellow drone of *koro* sounds.

Yesterday I gathered a few things—diapers, some small shirts, a bar of laundry soap, bathing soap, a box of talcum powder, a new bucket, and a small bundle of *yaqona*, and presented them to Momo. "'*Ere ere Momo Poi, luve Va me a caqu* (Please Uncle Poi, Va's baby, this is for her)," thus sanctifying the naming of the baby as my namesake and, in effect, taking her as my own. Momo says that when the baby is baptized, Malé and I will hold her. The new mat we will stand on will be given to the minister, and a small feast will be prepared for the family and the *mataquali*. Malé wants her to be ours. It would be nice to have a girl, a girl to care for and guide and, I hope, protect from the harshness of village life.

Malé and I often argue. His demands are continuous—massage my back, light my cigarette, bring me my tea, make my breakfast, get this, do that—until I explode in anger, and the argument is on. I waste myself in resistance, become absorbed in the conflict. Explanations are useless. I am his wife. Wives wait on their husbands.

I retreat to Wasa, climb up into the heavy wide branches of the *baka*, look around me at a world free of demands, at the purity of Seatura. In the wind-driven cool of the immense tree, my perspective returns, as it always does when I find the sanctuary of silence and nature. Below me are the family graves, the rough rock-walled partitions of ego, now liberated by death into something more gentle. The dead are not demanding. They do not ask where I have been and where I am going. Nothing asks. Just below me is Malé's great-grandfather's grave; an irregular wall of rocks and some sand is what is left of that intense demanding presence. He might have eaten men and women, beaten his wife, shouted at his children. He is silent now, companionable with the lapping tide, the soft scratching of sand crabs scuttling seaward. Behind me is Malé's great-uncle's grave. A rusty tin basin, an enameled cup, a water bottle—things he used in life—lie overturned on the grave. Nei's grave is covered with a now-shredding big *tapa*, of Togan style, once soft browns and creamy whites. There are names lettered across the edges, probably names of the women who made the *tapa*. It is slowly disintegrating in the rain and sun, back into the substance of air. On top of the *tapa* is an Ovaltine can filled with scummy water and three bent and torn branches of once fresh ferns. Next to the Ovaltine can is a plastic bottle containing the medicines given to Nei at the Labasa hospital, and in a glass jar are two vials of chemicals used to test her urine. I remember Va's bright face the day of the funeral when she said she was going to put these things on her mother's grave. Nice? she asked, and looked at me with big wet eyes. Two small freshly tended graves, planted with young ginger and *ti*, contain the memories of Vili and Banoko's first and third children, both dead at birth. Inside Botoi, Joana McIntyre Varawa stirs, utters a soft cry, and is hushed by Va's nipple.

Malé, sleeping behind me on the bed in our *bure*, talks to an unknown companion; his eyeballs twitch, he moans, and he is still. The wind flutters the *tapa* on the wall. The plastic beachball globe swings in the wind, then bounces. Continents twirl by. Bounce, bounce, bounce; now South America, now Africa, the vast expanse of the South Pacific, bounce around in the afternoon wind.

THE GRASS HOUSE: VEDRALA

Uncle Malé's brown, skinny body is folded into a perfect lotus position. He rolls, with intense concentration, a strip of leaf tobacco into a torn strip of newspaper, looks at his smoke with satisfaction, and lights it. He carefully puts a scrap of newsprint and a torn tobacco leaf back into a small tin box and looks up. Momo is kneeling. Behind him is a small bundle of *yaqona* and two folded mats. On the mat next to him is a *tabua*. A white pig is sleeping in the bush; she is part of the ceremony. Uncle Malé is wearing a printed *sulu*, faded, its cotton soft and clinging. The tiny ash in Uncle Malé's smoke glows. Malé is sitting next to me. It is afternoon, a cloudy day.

We are in a kind of play. Uncle Malé and Momo have talked about what each will do, the order of the ceremonies, what has been given, what will be given. Momo will *kerekere* the *vanua* (ask his older brother and all of their ancestors for this land for us on Vedrala). Uncle Malé will present us with the house site (*yavu*).

There is a bundle of *yaqona* on the mat on Uncle Malé's left side, a *tabua*, and a folded mat. Leaning on the fence behind him is a heavy hardwood post, wrapped with a green spiral of vine leaves. Viliame sits next to Uncle Malé. Small Malé comes and goes, sits properly, cross-legged, for a few moments, watching, then slides around Momo, then sits.

The day is cool. I am conscious of the greens, intrigued by the vine-wrapped heavy red post.

We are sitting in front of our small house on an old mat. We are about to begin to build a traditional *bure*, a permanent house for Malé and me. I am impressed with the plainness of everything.

Momo prays, calls for straightness, from the heart, from the deep inside center, straight from there to the God above (*Kalou mai lomalagi*), to the God below (*Alou mai bulu*), to the *vū*, to all that is buried under the soil, calls for it to be straight and true (*dodonu*) all the way from deep inside us—Malé, me, Momo, Viliame, Uncle Malé—all of us, to the God of the heavens above, to the spirits behind us, the spirits of ancestors, of those who are buried, that it all be cleaned.

He prays that the *yavu*, the foundation, the house site, will be made right, with all that has passed, with all that is given.

We give the *yaqona*, the *tabua*, the mats, the post, the pig, the precious things of this life, in this ceremony to sanctify this transfer of our family ground. Momo describes the *tabua*, pats the mats, touches the *yaqona*, prays. He looks across the years at his brother's face, sees both of them sitting here on their island, exchanging these precious things.

The mats are to clean away the spider webs from the rafters, to brush away the cobwebs, to clean the old house and to make comfortable the new house.

The heavy red post twined with bright green vines is to start the new house, to keep when we rebuild the house. That post is the gift of the house.

The pig is the sacred food that feeds us, that gives us life.

The *yaqona* is for the calm, for the grace, and for the time we have here, drinking together.

The *tabua*, tied with the *magimagi*, is to invoke our sacred connection, to bind this agreement.

The *yavu* is given.

Me dodonu ve'ea Kalou mai lomalagi
Me dodonu ve'ea Alou mai bulu
Me ua ni dua leqa
Me datou bula vata no ga
Me gai tawase
edatou ga a mate

mana

ai e dina

a

a vura

yea . . . yea.

Talo. Pour the stream of earth-tasting water back into the bowl, brown transparent water falling into the *tanoa*. Drinking, I look into the *bilo*, hold it with two cupped hands, let some *qona* spill out of the *bilo* onto the ground, then drink it all down. *Cobo*.

⯭ ⯭ ⯭

They carry the stones from the hillside, all who can carry. Small Suli, small Malé, the young boys, carry the stones, working to lift the weight, proud. Stones from the hill, sand dug in the moonlight by the sea, carried in bags, dumped in small piles on the growing *yavu*. Momo, Malé, Ratu, and Are dig the holes for the posts, lying flat on the ground to reach down for cupfuls of sand to make the holes deep enough. Measuring, eyeing, Momo walks back and forth with a stick, straightening, leveling, fastening, making our house.

I have taken picture after picture of the young boys tying the reeds together over the rafters. Pictures from inside—seeing them dark, competent, serious, tying the reeds with great attention and much joking. From the outside they look like extravagant birds, dangling thick bunches of split vines down their backs, along their legs. The boys are conscious of their elegance, weaving the frame, balanced on the rafters, the long tufts of vine feathers tucked into their waists. Momo's bunch—carrying stones, hauling bags of sand, carrying water, fishing, doing *meke* on the beach next to flaming heaps of dry coconut leaves—wonderful, goofy dark shapes in the firelight, shy and proud at the same time.

⯭ ⯭ ⯭

The rafters of the house curve up into the shadows above me, a soaring sculpture of polished branches. Some of the thatch is still green, pressed against the reeds. There is a small smoky fire burning in the center of the house, flaming in a depression scooped out of the sand. There is a pile of coconut husks to feed the fire, and the burning husks give off fragrant blue smoke, drifting up into the rafters: to dry the thatch, the *rau*; to clean and smoke the house. It is pleasant inside the house, warm and quiet. It has been raining, and it is good that now the house is finished. Just yesterday they finished the *rau*, tied off the peak of the golden thatch with hanks of heavy, rust red vines of the bush, beaten, softened into string. Two hollowed black trunks of tree ferns stand out from the ends of the twined roof, the setting sun burnishing the thatch, the velvety black trunks, dark slender extensions, balancing the house, completing it.

Now the roof is closed. The leveled sand of the *yavu* has been swept. The house is clean. There are piles of dry coconut leaves spread on the sand. The fire glows; blue smoke drifts into the rafters, warms and seasons the wood. I put a few more husks on the fire, pile them up properly, and sit back, looking upward, into the soft brown shadows.

Outside Malé and Ratu are pounding the grog. I hear the thunk of the *ta bili*, the laughter. Ratu's laugh rings out, catches everyone up in it, draws more laughter. Ratu is long legged, elegant. He is wearing brilliant orange overalls, his dark handsome face a flaring contrast of black eyes, hair, beard, and the glowing orange.

After much begging and *kerekere*, Ratu and Are will do a *meke* on the beach. Small Sione plays a makeshift *lali*, starts the beat. Malé is gleeful watching Ratu position himself on the sand. Ratu giggles, whoops, and breaks up in laughter. Malé howls, laughing with him, punches my arm, "Look, Joana, look at Ratu!"

Ratu strikes a pose, plants his fists on his knees, slumps, rolls his eyes, and looks at us sideways, clowning the gestures of the *meke*.

"*Meke! Meke!*" We *cobo*, call out "*Meke*!" to Ratu, who gets going with the motions. He is graceful and extravagant, making it all up, but perfectly. His body glows orange in the fading twilight; burning coconut fronds light the scene. Are and Ratu are thumping around in the *meke*, making spear motions and pounding motions and funny faces; and we are howling with glee, clapping, yelling, "*La'o, la'o!* Go, go, *meke, meke*!"

Later we go into the new *bure* to make the ceremony. All come inside. All. Momo, Uncle Tevita, Malé, Ratu, Are, Viliame, Bola, Taga, Ima, Vatea, small Malé, small Suli, Siti, Soro, Jone, Dabobo, Waisake, Sireli, and I—all who have lived together and eaten together and worked on the house for the last month are here, inside, to make the ceremony to finish the house.

Next to Momo are six 4-gallon drums of kerosene, a wrapped bundle of *yaqona*, a *tabua*. A *magiti* waits.

Momo speaks. Vedrala was empty, uninhabited (*lala*), and now it is filled. We have built a house together. We have done it properly; we have blessed each action. We have not had any trouble. No one has gotten hurt. We thank the God for the life, for the life he gives us. We thank the God for the health. We thank the God for the food we have to eat. We thank the God for the water we have to drink. We thank the God for this world he has given for us to live in.

We give you this kerosene so that you can see, so that you can have light. We give the *yaqona* for the love we have as a family, for one another, to take care of one another, to stay together.

We give this *tabua* for the *mana*, to bind and make true what we say here.

We ask that if we have done anything wrong, if we have left anything out, that we be forgiven, that it all be cleaned.

To make straight
    for the God of Heaven
to make straight
    for the God under the earth

that there not be any trouble
that we stay alive close together before we die
when we die the rest stay together
and we separate.

⁂

This book is for my mother, Lillian;
    my husband, Malé;
    my son, Ian;
and my namesake, Joana McIntyre Varawa.

NOTES ON CUSTOMS AND LANGUAGE

Not much longer ago than one hundred years Fiji was composed of many small tribes that were often at war with one another. The wars could be over almost anything, but they tended to isolate small groups from one another and intensify differences in speech and customs. Fijian culture developed a pattern that was recognizable throughout the group, but differed in details. Thus the words, the objects, the order, and the conduct of a ceremony might differ from place to place; but there would be a ceremony, and it would include the use of *yaqona*, the exchange of *tabua*, and respect for the conduct of the ceremony.

An exploration of Fijian customs would take more than another book, so the following is offered as the briefest sketch of some of the more obvious customs observed in Galoa that seem to be consistent throughout Fiji.

Respect is always accorded chiefs, but the forms will differ from place to place. Etiquette involving high chiefs is complex and very strict. Respect is also expected from women to men; from sons and daughters to fathers, then to mothers; from sisters to brothers; and from younger to older. Some places, particularly in Vanua Levu, have historic relative avoidance patterns that are more or less followed.

Throughout Fiji the head is considered *tabu*, or sacred, and not to be touched. Family members may touch one another's heads, but usually will ask permission by saying, "*Tilou* (excuse me)." When taking something from above someone's head—from a shelf or wall, for example—the customary form is to say *tilou*, take the object, then sit next to the person and clap three times with cupped hands in the *cobo*. One should not stand above another person, but should sit at the same level. This is particularly true in the presence of chiefs. Respect calls for stooping or crawling in and out of houses in the presence of others. When passing in front of or behind a person, one should say *tilou*. *Yaqona* drinking is accompanied by a complex etiquette, and ceremonies involving chiefs are very strict. It is *tabu* to cross the *magimagi* cord attached to the *tanoa* while serving *yaqona*, but in Macuata it is crossed deliberately as part of the ceremony.

Men are expected to sit cross-legged while eating or drinking *yaqona*, and women sit with their legs tucked under them. Small children are allowed a great deal of leeway, but are gently taught to observe the proper forms as soon as they are old enough. Silence in adults is valued and considered a form of respect and character. Soft speaking is also valued.

Very often men eat first and women later. When eating together the men usually sit with men and the women with women. Small children are permitted to range freely, even to walk across the tablecloth. Table manners are not impressed on young children. Food is most often eaten with the hands, and the hands are washed in a small bowl of fresh water after eating.

Women are expected to behave modestly at all times and to keep their lower bodies well covered. Men, on the other hand, often play and tease in a sexual mode during informal *yaqona* sessions. Women, among themselves, often joke and mimic men's activities. Nursing mothers and old women may uncover their breasts without comment. Men and women do not touch one another in public or in view of the family.

Toilet training in young children is casual. When they are old enough they are sent outside to the bush or beach to relieve themselves; rarely do young children use outhouses or toilet enclosures. Children are not scolded when they dirty their clothes or persons in play. Personal cleanliness is valued, but it is expected that people will get dirty while working and playing.

Cleanliness and propriety are demanded during church service and on Sundays. Sunday is observed from twelve midnight on Saturday to twelve midnight on Sunday. No loud talking, playing, working (except for food preparation) is permitted during those hours. Singing is confined to hymns, and radios are kept quiet or turned to religious programs. *Yaqona* is drunk on Sundays.

In general, Fijian custom values skill, hard work, mannerliness, observing the proper forms, and respect. People who live closely together through generation after generation must get along, and the good Fijian is the one who makes it easy for the group to function. Egotism, personal display, and competitiveness are disdained. Humor and pleasantness are encouraged, and often the social skids are greased with laughter.

⋆ ⋆ ⋆

The Fijian language is said to belong to the Malayo-Polynesian language group. I found many words similar to those I know from Hawaiian, although my knowledge of Hawaiian is extremely limited.

Understanding the language is confounded by a multiplicity of dialects, which reflects the localized tribal state of the culture. Even very commonly used words, such as *yaqona*, will undergo spelling changes. The dictionary, *A New Fijian Dictionary*, compiled by A. Capell for the government of Fiji, lists fifteen forms of cardinal pronouns for fifteen different dialectic groups, and that list includes only the major dialectic groups. There are three forms of possessives: general possession, possessing something to eat, possessing something to drink. All of these forms change from dialect to dialect. The complexity of the language lies in the direction of specifying relationships of action: who will go with whom and who will stay; who is included and who is not; descriptions of motion, the natural environment, the materials of the culture. Humor is based on word play or punning on possible meanings of the same word.

The official language of the nation is Bauan, adopted because the high chief of Bau was the conquering ruler of much of Fiji at the time of cession. The language spoken on Galoa is a distinct dialect, which is not even shared by the small neighboring islands of Tavea and Yaqaga and the neighboring villages of the mainland. Galoan is distinguished by the dropping of the *k*, substituting instead a glottal stop, indicated in writing by an apostrophe. Thus the word for eat, *kana*, in Bauan becomes *'ana* in Galoan; *lako*, to go, which is used for motion of any kind, is *la'o* in Galoan. In addition to the dropping of the *k*, Galoan has many words distinctly its own: no, for example, which is *sega* in Bauan, is *maqa* in Galoan; drink is *gunu* in Bauan and *somi* in Galoan.

Children are taught in the Bauan dialect at school; thus it is possible for most Fijians to communicate across dialectic lines. Written English is taught as a second language, beginning at the first-grade level. Thus a native speaker will know his or her own dialect, Bauan, perhaps some English, and perhaps some Hindi, which is also taught to Indian-speaking children. Indian-speaking children, who often attend separate Indian schools, are taught Hindi, then Fijian and English.

The definitions that follow are based on my own imperfect understanding of the Galoan dialect and may not agree with those offered by a native speaker. I have tried, as much as possible, to check definitions with family members. Also, as mentioned in the text, the language is elusive because of the Fijian habit of assigning many subtle meanings to the same word, meanings which rely on usage and on metaphor.

Although I have tried to define Fijian words as they appear in the text and to identify places and family members in the narrative, the following glossary is offered as a reference to aid clarity.

Pronunciation is fairly simple, with the exception of a few letters that are written to convey double consonants. Fijian is always spoken with a vowel-consonant combination, yet there are certain consonants that have double sounds. Early missionary orthographers devised the following system for writing double consonants, for Fijian was not a written language until the coming of the missionaries.

Double consonants are pronounced as follows:

b represents *mb*, as in member, bula = mbula

d represents *nd*, as in find, Nadi = Nandi

q represents *nng*, as in finger, Qare = Nngare

g represents *ng*, as in sing, sega = senga

c represents *th*, as in then, cobo = thombo

Vowels, in general, are pronounced as in Italian and Spanish. Consonants are pronounced as in English.

GLOSSARY

a cava what? pronoun form of *cava*, what, which.

aisau a small nectar-eating bird.

'ana va levu eat a lot, Galoan dialect.

au sa marau I am happy.

avia tree with crisp red fruit that looks like a small apple.

a vūra a phrase used in the *yaqona* ceremony, meaning obscure, may come from the verb *vura*, to arrive, emerge, appear.

baka *Ficus obliqua*, Moraceae, the banyan tree, considered by some to be the residence of evil spirits.

bati a tooth; *bati balavu*, long tooth, name of historic *mataquali* on Galoa; *bati lekaleka*, short tooth, *mataquali* on Galoa; *bati ni namu*, the tooth of the mosquito, a stone on Galoa considered sacred.

Bau a small chiefly island off the south coast of Viti Levu which once ruled most of Fiji, source of the Bauan dialect, the official language of the group.

bele *Hibiscus Manihot*, Malvaceae, an edible green.

Beqa an island off Viti Levu, home of *mataquali* that has the power to walk the burning stones of the *lovo* without harm.

bia cassava, Galoan dialect.

bilo cup, usually made from the coconut shell.

boci penis.

bokola a dead body for eating.

bola woven coconut-leaf mat used in house thatching.

bonito a large deep-water fish, Galoan dialect.

Botoi name of Varawa family *bure*.

Bua province in Vanua Levu, north coast; *bua*, *Fagraea Berterana*, Loganiaceae, a native tree.

bula health, life, a greeting; *bula re*, good health, Galoan dialect.

bure a Fijian sleeping house.

burua a slaughtered animal divided as food for mourners at a funeral.

cagi wind; *cagi bula*, fresh or trade wind.

cago *Curcuma longa*, Zingiberaceae, turmeric.

Cakau ni Tabua reef of the *tabua*, a reef off the coast of Bua once heaped with the bones and teeth of dead whales.

cassava *Manihot utilissima*, Euphorbiaceae, the tapioca plant, of which both roots and leaves are edible.

cicicere a ceremony in honor of a new boat, to run in the low tide.

cina a lamp or torch.

cobo to clap with cupped hands.

coco'a fishing in the shallow water with a hand-thrown spear, Galoan dialect.

copra dried coconut meat.

cori bola to tie coconut mats to a house frame.

dadakulaci a black-and-white banded sea snake that is said to be poisonous, the *vū* of the Varawa family *mataquali*.

dalo *Colocasia esculenta*, Araceae, a widely used edible plant, both leaves and roots are eaten, known as *taro* in much of Polynesia.

davui a large triton shell used as a trumpet.

dela on top of; *dela ni Seatura*, the summit of a mountain in Bua; *dela ni yavu*, the top of the house foundation, refers to where one comes from, where one was born.

dhal dried yellow peas, boiled and served as soup, introduced by East Indians into Fiji.

dodonu correct, straight, right.

dou you three, you few; *dou cobo*, you clap; *dou somi a veiqaravi*, you who have prepared the *yaqona* drink.

Drana a swampy area in the mountains above Lekutu where *dalo* is planted.

dri Holothurioidea, the edible sea cucumber, *bêche-de-mer*.

duo duo traditional respectful way of announcing one's presence during the day, Bua province.

e in, at, on; *e vi*, where?

ē ō dua sa dua sa, ā muduō a phrase used in the *yaqona* ceremony, meaning obscure, may come from the verb *mudu*, ceased, ended, cut off; thus Ah! It is over!

frangipani *Plumeria acutifolia*, Apocynaceae, an introduced tree with fragrant flowers, sometimes called the *bua vavalagi*, the foreign bua.

Galoa a small island off the north coast of Vanua Levu; the name may refer to *gā loa*, the black duck.

gasau a reed used to thatch houses.

gone a child of either sex; *gone dau*, *mataquali* of expert fishermen.

guto wood used for cooking, a burning brand used for lighting fires or cigarettes.

i'a fish of any kind, Galoan dialect.

īo yes.

isa an expression of yearning, regret, remembrance.

isalei *isa* intensified.

jaba see *sulu*.

Kalou the Christian God, any god; *kalougata*, God's blessing.

kavuru to share cigarettes.

kerekere to beg or plead for, usually from family relatives; *'ere'ere*, Galoan dialect.

koa (Hawaiian) *Acacia Koa*, leguminosae, a small scrubby coastal tree with feathery leaves.

koro a village.

kosa the dregs of the *yaqona* root after the first mixing, sometimes repounded and infused again.

Labasa the principal town and trading center in Vanua Levu.

lailai small; used to distinguish younger children from their older namesakes; a little.

lali traditional hollowed-log drum beaten with two sticks.

la'o to go, used for motion of any kind; *la'o mai*, come here, Galoan dialect.

Lau a string of islands to the south of Viti Levu, a province, heavily influenced by Tongan culture.

Lekutu the name of a village near Nakadrudru, on the banks of the Lekutu River.

leqa trouble or problems of any kind; *leqa levu*, serious problem; *maqa leqa*, no problems.
levu large, big, great, much.
lialia foolish, crazy, silly.
liku formerly a short skirt of roots or pounded bark, now any kind of a skirt.
lolo coconut cream made by squeezing the grated coconut meat.
loloma love, pity, mercy, sympathy.
lotu prayer, a Christian; *Lotu*, the Christian church.
lovo oven made by digging a hole in the earth, filling it with wood, and placing a layer of stones on top of the wood; when the fire burns down, the stones are red hot and the food is placed on the stones, covered with leaves and earth, and left for several hours to steam.
Macuata a province on the north coast of Vanua Levu.
madua ashamed, shy.
magimagi traditional cord or rope woven from the beaten fiber of the coconut husk.
magiti a feast, cooked food.
mai 'ana come and eat, Galoan dialect.
Malau a large lumber mill near Labasa.
malumulumu weak, soft, gentle.
mana supernatural power, extraordinary ability.
maqa no, not.
marama a woman.
mata the eye, the face, the source or front of a thing; *mata ni gasau*, a ceremony to clear the way of danger and to offer atonement; *mata ni gone*, a ceremony held the first time a child is brought to his family's village; *mata ni vanua*, the chief's herald or spokesman.
mātaisau skilled carpenter, a class of skilled carpenters, a *mataquali*.
mataquali the extended family, the landowning unit; formerly a clan distinguished by special hereditary skills and responsibilities.
mate death, disease, to die.
mautu a way things are done carefully and well.
meke the traditional storytelling dance and song.

me ua ni dua leqa that there not be a single trouble.

moce sleep, goodnight, goodby.

moli sweet tasting, orange, lemon; an expression of thanks after drinking *yaqona* or to acknowledge a good wish.

momo uncle.

mosi pain.

moto a spear for fishing.

muni belonging to everyone; *muni savusavu*, the sandspit in Galoa.

na mother, the articles a, the.

Nabau a freshwater artesian well on Galoa.

Nabouwalu a settlement in Bua province, provincial seat of government and law.

Nadi a large town in western Viti Levu, international airport.

Naicobocobo a point of land on Bua Bay where the spirits of the dead leap into their future life, a *tabu* place.

Naigani an island off the coast of Bua where the bones of the *saqa*, thrown into the sea, become *saqa* again.

Naivaka a coastal village in Bua opposite Yaqaga Island.

Nakadrudru a settlement near Lekutu.

Nalomolomo family garden plot on Galoa, name of *mataquali*.

Namoa a tiny island off Vatoa reef.

Namoi family garden land.

Naniqaniqa a small island off the coast of Bua.

Na Vanua Yalewa the island of the women.

nei aunty.

noqu yaca my name, term of address for one's namesake.

nunu to dive under the water.

oilei expression of intense grief, formal cry of mourning.

oqo here, near the speaker.

pāpāpā gossip or chatter, a word invented by Jone Varawa.

pua hilahila (Hawaiian) a ground-hugging sensitive plant.

qona see *yaqona*.

rara *Erythrina variegata* var. *orientalis*, Leguminose, a deciduous tree with brilliant orange flowers.

rau grass or reeds for thatching a house.

rere to be afraid, fearful.

Rewa the heavily populated delta around the great Rewa River, near Suva in Viti Levu.

roaroa morning, in the morning, a goodnight salutation, Galoan dialect.

Ro Da Cewa the name of the *mataquali* land where *bati ni namu* is located, Galoa.

roti East Indian flat pancake made of unleavened flour.

sa adds emphasis; *sa mada yaqona va turaga*, the *yaqona* for the chief is empty; *sa oti*, it's finished; *sa rauta*, it's enough; *sa re*, it's good.

salusalu traditional garland of flowers woven with streamers of *tapa*.

saqa a large fish, the trevally, *Caranx* species.

sari an East Indian garment made by wrapping several yards of cloth around the body and over the shoulder.

sāsā dried coconut leaves used as flooring under mats, a broom made from the midrib of the coconut leaves, any dried coconut leaves.

Savusavu a town in Cakaudrove province, the provincial seat of government, a tourist center.

Seatura a large volcanic shield mountain that comprises most of Bua province.

sele to cut, any knife; *sele levu*, a large machetelike knife used for gardening, house building, wood cutting, food gathering and preparing; the all-purpose knife of the Fijian villager.

sevusevu a presentation of *yaqona* for a ceremony of returning.

siga day, daylight, sun; *Siga Tabu*, Sunday.

sili to bathe, wash the body; *vale ni sili*, a bathing house or enclosure.

soli to give, usually to the church or community-supported institution; a gift, offering, collection, tax.

somi to drink; *somi qona*, drink yaqona; *somi ti*, drink tea, Galoan dialect.

Somosomo chiefly village on Taveuni, Cakaudrove province.

soro an offering to obtain forgiveness, usually of *tabua* and *yaqona*.

suki native leaf tobacco.

sulu two-meter length of brightly printed cloth wrapped around the waist and worn by both men and women; *sulu* and *jaba*, straight sewn long skirt with fitted tunic-length top worn by women; *sulu va taga*, knee-length sewn wrapped *sulu* with pockets worn by men.
susuna a tree related to the pandanus, the leaves of which are used for rolling leaf tobacco.
Suva principal city and capital of Fiji.
ta father.
ta bili mortar used for pounding *yaqona* roots.
tabu sacred, forbidden.
tabua a whale's tooth, both ends of which are tied together by a string, usually of *magimagi*, used in every important ceremony—birth, death, betrothal, marriage, land transfer, to seek forgiveness, to express honor and respect, and so on—the most sacred object in Fijian culture.
talanoa a conversation, storytelling.
talo to pour, to serve a round of *yaqona*; *talo qona va turaga*, to serve a round of *yaqona* for the chief.
tama a traditional respectful call to announce one's presence; a different call is used during the day and night, and by men and women; in Bua province used by everyone; in other places used only toward chiefs.
tamata a human being.
tanoa the traditional four-legged wooden bowl for serving *yaqona*.
tapa (Hawaiian) a paperlike cloth made from the pounded bark of the paper mulberry tree and decorated with intricate stenciled designs; the Fijian term is *masi*, referring to the name of the tree.
taro (Hawaiian) the *dalo* plant.
tate a small container used for dipping up water.
taukei the landowner.
tavale a first cousin.
Tavea a sister island to Galoa off the coast of Vanua Levu.
Taveuni a large island southeast of Vanua Levu.
teitei a garden, to work in the garden, to plant or dig.
tēvoro an evil spirit, a demon; maybe a recall of ancient gods now considered demonic by Christians.

ti the English word *tea*.
Tilaro name of family boat, a land crab.
tivi the Fijian chestnut tree, formerly sacred.
tobu a pit or well; *tobu ni madrai*, a pit dug in the ground to store fermenting cassava or breadfruit to be used in the making of native bread.
tolo the oldest member of a *mataquali*.
Tonga the Kingdom of Tonga, a string of islands to the southeast of Fiji; the source of much cultural influence.
totolo to hurry.
tovako tobacco.
tubea to hold in the hand; *tubea qona va turaga*, gather up the *qona* for the chief preparatory to mixing it.
turaga a chief, a lord, sometimes used in prayer to refer to God; *sau turaga*, a *mataquali* whose traditional role was to select the chief from possible members of the chiefly family.
tuva a vine used for lashing in house building, Galoan dialect.
ulu the head, the hair, the breadfruit tree and fruit; *ulu matua*, the first-born child.
uvi *Dioscorea alata*, Dioscoreaceae, a yam.
vaivai name used to describe several different trees that have in common small leguminous leaves, Galoan dialect.
vaka Viti the Fijian way.
vale any kind of a house; *vale ni mate*, the house of mourning where the women sit in wake; *vale ni sili*, bathing house; *vale ni uro*, cooking house; *vale vō*, toilet or outhouse; *vale kau*, wooden house.
vanua land, region, place; a physical and mystical union of land, sea, and chief; Vanua Levu, big land, the name of Fiji's second-largest island.
vasua a large saltwater rock-boring clam.
vata a shelf or platform.
Vatoa an extensive reef in Bua.
vatu a stone or rock; *vatu nu loa*, a temporary shelter of thatched roof set on poles, or any temporary roofed shelter.

vau *Hibiscus tiliaceus*, Malvaceae, a tree; the bark is used to make string for house lashings, tying firewood, gathering fish, and so on.

Vedrala a small island near Galoa.

Verata said to be the original landing place of the ancient Fijians, a village on the coast of Viti Levu.

vinaka good, thank you; *vina'a va'a levu*, thank you very much, Galoan dialect.

Viti Levu largest island in Fiji, means big Fiji.

voivoi *Pandanus Thurstonii*, Pandanaceae, a kind of pandanus; the leaves are used to make mats, baskets, fans, and, formerly, sails.

volaca a prized reef fish.

vonu a sea turtle, several different varieties, considered the suitable food for ceremonial occasions, and required for ceremonies involving chiefs.

vosota to endure, be patient, endure suffering.

vū basis, root, ancestor, bottom, an ancestral animal, the eating of which is *tabu*.

vuaka a pig, the ceremonial food of the land, essential in ceremonies.

vuci a wetland *dalo* garden.

vuvu to share cigarettes.

wai water, liquid of any kind.

Wainunu a large bay and surrounding land area in Vanua Levu.

waka the root of a plant, the *yaqona* root.

walu a kingfish, the number eight.

waqa a large boat, a box, container, coffin, formerly a canoe.

Wasa family burial ground on Galoa.

wawa wait; *wa*, to wait.

yadra to wake up, good morning, a greeting.

yalo spirit, soul, often travels while the body is asleep.

Yaqaga a large island on the north coast of Vanua Levu.

yaqona *Piper methysticum*, Piperaceae, a plant with tranquilizing properties, known as *kava* throughout the South Pacific, widely used; the roots are dried, pounded, and infused with water; name refers to entire plant, roots, and resulting drink; the drinking of *yaqona* was formerly restricted to chiefs or ceremonial use; now a social drink, primarily among men.

Yasawa a string of islands north of Viti Levu.

yavu the foundation of a house, the place one comes from, the inherited and presented house site in the village.

ABOUT THE AUTHOR

Joana McIntyre Varawa was born in Los Angeles and educated at UCLA and Berkeley. She was public affairs program producer at Pacifica Radio and special projects coordinator for Friends of the Earth, where she originated a national media campaign to make it unfashionable to wear the furs of wild and endangered animals. Varawa was the founding president of Project Jonah, the international organization to save the whales. She is the author of two previous books, *Mind in the Waters* and *The Delicate Art of Whale Watching*. Varawa has one son, who lives in America; she and her husband Malé live on the island of Vedrala, near Galoa, in Fiji.